INFRARED AND SUBMILLIMETER ASTRONOMY

ASTROPHYSICS AND SPACE SCIENCE LIBRARY

A SERIES OF BOOKS ON THE RECENT DEVELOPMENTS
OF SPACE SCIENCE AND OF GENERAL GEOPHYSICS AND ASTROPHYSICS
PUBLISHED IN CONNECTION WITH THE JOURNAL
SPACE SCIENCE REVIEWS

VOLUME 63
PROCEEDINGS

INFRARED AND SUBMILLIMETER ASTRONOMY

PROCEEDINGS OF A SYMPOSIUM HELD IN PHILADELPHIA, PENN., U.S.A., JUNE 8-10, 1976

Edited by

GIOVANNI G. FAZIO

Center for Astrophysics, Cambridge, Mass., U.S.A.

Sponsored by

COMMITTEE ON SPACE RESEARCH (COSPAR)
INTERNATIONAL ASTRONOMICAL UNION (IAU)
INTERNATIONAL UNION OF RADIO SCIENCE (URSI)

D. REIDEL PUBLISHING COMPANY

DORDRECHT-HOLLAND/BOSTON-U.S.A.

ISBN 90-277-0791-X

Published by D. Reidel Publishing Company,
P.O. Box 17, Dordrecht, Holland

Sold and distributed in the U.S.A., Canada and Mexico
by D. Reidel Publishing Company, Inc.
Lincoln Building, 160 Old Derby Street, Hingham,
Mass. 02043, U.S.A.

Printed in The Netherlands

TABLE OF CONTENTS

PREFACE

The Symposium on Infrared and Submillimeter Astronomy was held in Philadelphia, Pennsylvania, U.S.A., on June 8-10, 1976, as an activity associated with the Nineteenth Plenary Meeting of the Committee on Space Research (COSPAR). The Symposium was sponsored jointly by COSPAR, the International Astronomical Union (IAU) and the International Union of Radio Science (URSI).

COSPAR is an interdisciplinary scientific organization, established by the International Council of Scientific Unions in 1958, to, in the words of its charter, "provide the world scientific community with the means whereby it may exploit the possibilities of satellites and space probes of all kinds for scientific purposes and exchange the resulting data on a co-operative basis." The purpose of this particular COSPAR Symposium was to present new results in infrared and submillimeter astronomy obtained by observations on aircraft, high altitude balloons, rockets, satellites, and space probes. Topics discussed included the Sun, the solar system, galactic and extra-galactic objects as well as the cosmic background radiation. Instrumentation for observations in infrared and submillimeter astronomy was also discussed, with particular emphasis on future programs from space observatories.

This particular Symposium was unique in many ways. It is the first symposium on the subject of infrared and submillimeter astronomy to be held at a COSPAR meeting. It was also one of the rare occasions in which an international group of infrared astronomers, primarily interested in space astronomy, has gathered together to exchange ideas and information. The number of astronomers that attended the Symposium is another indication of the rapid growth and importance of space astronomy in the infrared and submillimeter region of the spectrum.

This Symposium originated in Working Group 3 of COSPAR. This Working Group, chaired by Dr. Z. Švestka, is concerned with Space Techniques as Applied to Astrophysical Problems. I am indebted to the Program Committee, consisting of Drs. A.

Barrett (URSI); Dr. H. Friedman (IAU and COSPAR); Dr. Z. Švestka (COSPAR); Dr. H. Fechtig (COSPAR); Professor S. Hayakawa (COSPAR); and Professor S.L. Mandelshtam (COSPAR) for their guidance and assistance. I would also like to thank the session chairpersons and the invited speakers for their contributions to this Symposium. I would also like to express my sincerest gratitude to Professor C. de Jager, President of COSPAR, to the COSPAR Secretariat, in particular Mr. Z. Niemirowiz, and the local Organizing Committee, for arranging the many administrative details and for providing excellent facilities for the lectures.

Due to limitations on the number of printed pages, only the invited lectures are presented as complete papers. The contributed papers are listed and summarized in abstracts. In some cases the invited papers were not available and only an abstract is quoted. The papers were prepared by the authors themselves and submitted in camera-ready form. In some cases minor textual changes were made to correct errors. I wish to thank Dr. A.C. Strickland for her excellent and rapid editing of the papers and Mrs. Helen Beattie for assisting with the typing.

Giovanni G. Fazio
Chairman
Program Committee

PART I

INFRARED ASTRONOMY (GENERAL)

INFRARED SPACE ASTRONOMY--AN OVERVIEW

Frank J. Low
Lunar and Planetary Laboratory
University of Arizona
Tucson, Arizona 85721

I. INTRODUCTION

During the last 12 years infrared astronomy has grown from its infancy, when only the bright near-by planets could be studied, to a level of maturity permitting extensive observation dealing with some of the most important unsolved problems of modern astrophysics. This rapid growth to young adulthood was made possible by combining advances in instrumentation, such as the development of helium cooled detectors, low background telescopes and multiplex spectrometers, with high altitude and space platforms, such as jet aircraft, balloons and rockets. At this meeting we have the exciting opportunity to discuss major programs of the future such as IRAS (the joint Dutch--U.K.--U.S.A. infrared satellite project) and the equally exciting prospect of reviewing current progress in a large number of on going projects. Thus we should be able to see where the field stands now and where it is headed.

In this introductory paper it is not possible to give a comprehensive overview of all the excellent work now taking place in infrared space astronomy. Therefore, only an outline will be presented of how ground-based, airborne, balloon and rocket programs have combined to reveal the panarama of infrared sources which make up our current picture of the infrared sky. This picture of the infrared sky, as we see it today, will be illustrated by choosing for discussion outstanding examples of several classes of distinctly different types of sources. It is interesting to observe that as a result of this conference we may need to revise upward the finite, but steadily increasing, number of different designations needed to classify all the known infrared sources.

G. G. Fazio (ed.), Infrared and Submillimeter Astronomy, 3-11.

II. OUTLINE OF IMPORTANT INSTRUMENTATION

All of the observational results presented at this conference will have been made by (1) groundbased telescopes ranging in size from 0.1 to 5 m (2) airborne telescopes from 0.3 to 0.91 m (3) balloon telescopes from 0.02 to 1.0 m or (4) rocket telescopes about 0.15 m in size. There are distinct advantages and limitations for all the various instruments now in use. One reason there are so many different instruments is that they complement each other to such an extent that no one type of instrument can satisfactorily replace another.

As an example, there are continuing far-infrared observations with both large and small airborne and balloon telescopes. The airborne facilities resemble ground-based observatories and can provide high pointing accuracy (arcseconds), flexibility (complex focal-plane instruments easily accommodated) and reliability (rapid repair and turnaround following a malfunction). The balloon instruments provide lower background levels, much less pressure broadening of telluric lines, and longer more stable observing conditions; however, balloon instruments more closely resemble satellites than ordinary observatories. Rockets, of course, achieve much lower background levels but provide only short observation time and are costly. There is no doubt that the ultimate infrared telescope must be placed in orbit to combine long observation time with the lowest possible background levels to achieve the highest possible sensitivity. Even so it must be noted that angular resolution is frequently more important than sensitivity and large groundbased telescopes working through atmospheric windows provide the highest resolution at present.

The groundbased spatial interferometers that are just now being developed will soon extend angular resolution well beyond present limits and, when ultimately deployed, will greatly extend the resolution obtainable from space platforms.

III. THE INFRARED SKY

In order to easily construct an illustrative map of the infrared sky, the following somewhat arbitrary rules were adopted: (1) It was decided to exclude the bodies of the solar system, realizing that this important area would be covered later in the conference. (2) Emphasis would be placed on representing as many different types or classes of sources as can be _clearly_ discerned with _existing_ data. (3) Each class of object _would_, in general, be represented by its "brightest members".

Figure 1 contains an assemblage of objects chosen according to the above rules. Clearly there are mistakes and oversights

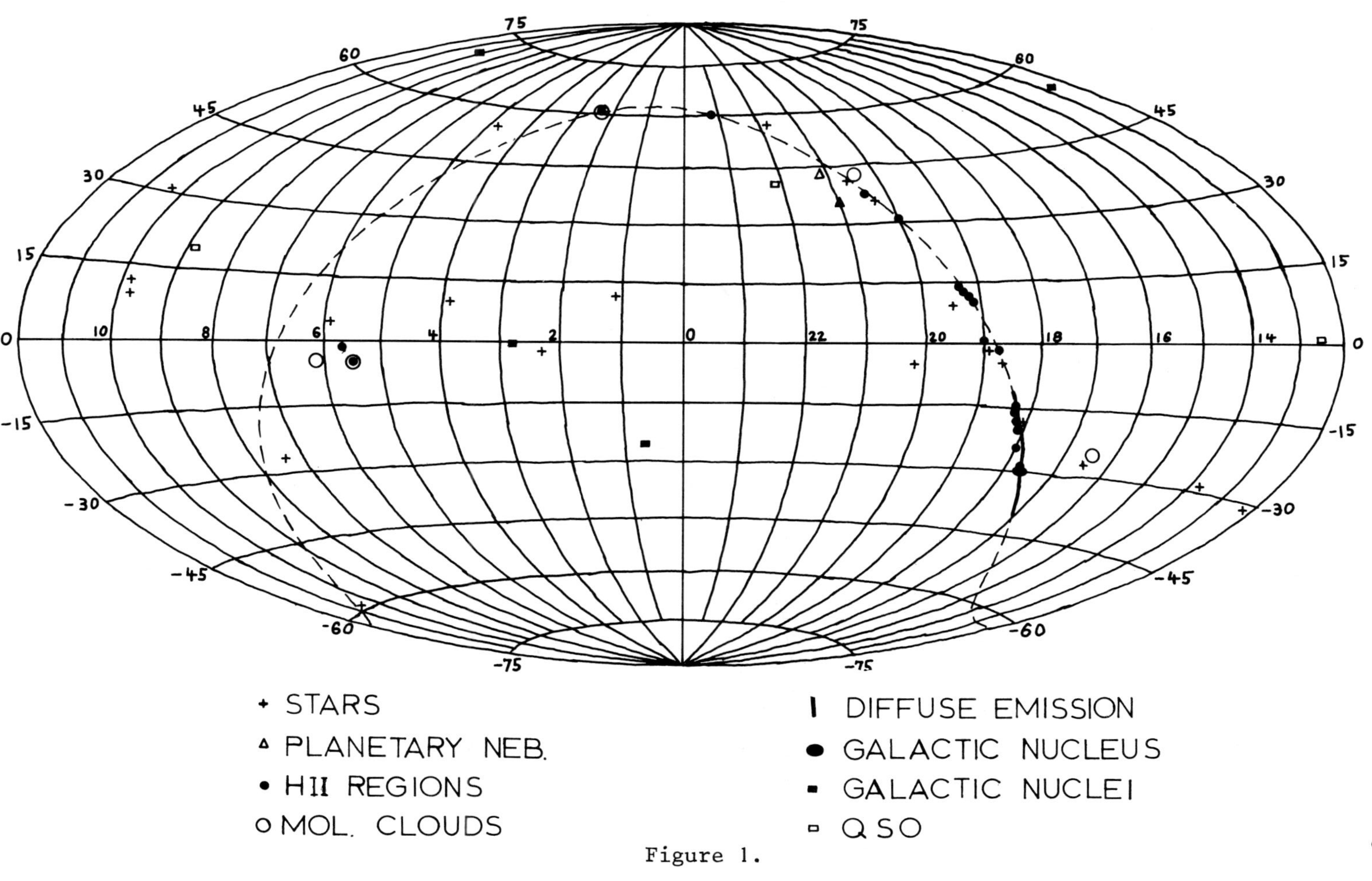
STARS
PLANETARY NEB.
HII REGIONS
MOL. CLOUDS
DIFFUSE EMISSION
GALACTIC NUCLEUS
GALACTIC NUCLEI
QSO

Figure 1.

as well as over-simplifications. Only a glance reveals the fact that the Northern sky is much more fully explored than the Southern sky. It is easy to pick out the galactic plane and certain prominent features of the plane such as the galactic center region.

After consultation with several colleagues, the eight categories or classes listed in Table 1 were chosen. The sources listed in each class are plotted with an identifying symbol in Figure 1. Note the extremely large range of apparent brightness between the various sources (∿0.04Jy for QSO to ∿10^6Jy for the brightest galactic sources). The following discussion of each class will clarify their definitions and, to some extent, their interrelations.

1. "Stars"

Only the 20 brightest stars in the AFCRL catalogue at 20 μ have been plotted along with ηCar, the brightest of all. This broad classification includes a wide range of objects. These are (1) ordinary stars emitting at a few thousand degrees (2) highly evolved stars of several types which eject sufficient mass to support circumstellar envelopes, emitting at temperatures from 100°K to as high as 1200°K (3) stars such as novae which eject mass on a less continuous basis but which nevertheless form circumstellar dust in great quantities (4) young stars or possibly "protostars" which are still embedded in their "protostellar" or "preplanetary" envelopes.

Dust laden envelopes or shells are found emitting over an extremely wide range of temperatures (1200°K > T > 100°K), however, it is common to find most of the emission peaked at 5 μ (∿500°K). Because of their distinctive spectral features at 10 and 20 μ "silicate" type particles are widely detected although other compositions must also be abundant. A surprisingly wide range of stellar luminosities are found despite obvious observational selection effects. Frequently, infrared stars are joined in small clusters and are associated with molecular clouds and HII regions. Extreme reddening sometimes plays the dominant role in fashioning the observed spectrum of infrared stars.

2. Planetary Nebulae and the "Egg Neb."

Compact planetary nebulae such as NGC7027 emit an infrared continuum at about 250°K. This phenomenon, where dust co-exists with ionized gas, is similar to that found in HII regions except that the dust temperatures tend to be higher. The first infrared emission lines were found in planetary nebulae. CRL2688, dubbed

Table 1. The Brightest Members of the Eight Presently Known Classes of Infrared Sources

1. Stars[4]

AFCRL	α(1950)	δ(1950)	20μ Flux Den.(Jy)	Name
157	1 04	12 19	0.83×10^3	CIT 3
318	2 17	- 3 12	1.9	o Cet
529	3 51	11 14	1.3	NML Tau
664	4 57	56 07	1.0	TY Cam
836	5 52	7 25	2.1	α Ori
1111	7 21	-25 41	11	VY C Ma
1380	9 45	11 39	1.1	R Leo
1381	9 45	13 30	∿25	IRC 10 216
1403	10 13	30 49	0.83	RW LMi
--	10 43	-59 25	∿50	η Car
1531	12 01	-32 04	2.7	IRC -30187
1650	13 46	-28 07	1.9	W Hya
1863	16 26	-26 19	1.0	α Sco
2071	18 05	-22 16	2.1	VY Sgr
2210	18 36	- 6 51	2.5	EW Sct
2251	18 45	- 2 03	2.5	AB Agl
2390	19 24	11 16	3.3	IRC+10420
2514	20 08	- 6 25	1.4	IRC -10529
2560	20 20	37 22	1.1	BC Cyg
2650	20 45	39 56	4.8	NML Cyg
2802	21 42	58 33	0.69	μ Cep

2. Planetary Nebulae

AFCRL	α(1950)	δ(1950)	20μ Flux Den.(Jy)	Name
2688	21 00	36 30	2.5×10^3	"Egg Neb."
2713	21 05	42 02	0.6	NGC7027

3. HII Regions

AFCRL	α(1950)	δ(1950)	20μ Flux Den.(Jy)	Name
326	2 22	61 52	5.6×10^3	W3
779	5 33	- 5 27	5.9	M42
807	5 39	- 1 57	3.3	NGC2024
2006	17 44	-28 33	0.7	Sgr B2
2052	18 01	-24 21	3.3	M8
2078	18 06	-20 19	3.0	W31

3. HII Regions (Continued)

AFCRL	α(1950)	δ(1950)	20μ Flux Den.(Jy)	Name
2090	18 11	-17 58	1.4×10^3	HFE 50
2117	18 16	-13 46	1.4	M16
2124	18 18	-16 13	20	M17
2245	18 43	- 2 42	1.1	HFE56
2304	18 59	1 08	1.0	W 48
2334	19 08	9 02	1.7	HFE 58
2341	19 11	10 48	1.2	?
2376	19 20	13 59	1.2	HFE59
2381	19 21	14 24	1.7	HFE 60
2495	20 00	33 25	1.4	NGC6857
2584	20 26	37 13	1.2	Sharp 106
3048	23 12	61 12	3.6	NGC 7538

4. Molecular Clouds

AFCRL	α(1950)	δ(1950)	100μ Flux Den.(Jy)	Name
326	2 22	61 52	$\sim 10^4$	W3 IRS 5
779	5 33	- 5 27	1×10^5	KL Neb
877	6 05	- 6 23	5×10^4	Mon R2
--	16 23	-24 17	3×10^4	ρ Oph DK.Cl.
--	20 37	42 12	$\sim 10^4$	W75 S OH

6. Galactic Nucleus

AFCRL	α(1950)	δ(1950)	20μ Flux Den.(Jy)	Name
2003	17 43	-28 54	2.6×10^3	Sgr A

7. Galactic Nuclei

AFCRL	α(1950)	δ(1950)	20μ Flux Den.(Jy)	Name
--	00 45	-25 34	30	NGC 253
--	2 40	00 20	60	NGC 1068
1388	9 52	69 55	100	M82
--	12 55	56 15	6	MK231

8. QSO

AFCRL	α(1950)	δ(1950)	10μ Flux Den.(Jy)	Name
--	08 53	20 15	$0.04 \rightarrow 0.07$	OJ 287
--	12 27	2 20	$0.2 \rightarrow 0.5$	3C 273
--	22 01	42 12	$0.2 \rightarrow 0.7$	BL Lac

the "Egg Nebula" is thought to represent the missing link between stars and planetary nebulae.

3. HII Regions

The 20 brightest HII regions at 20 μ listed in the AFCRL catalogue are plotted in Figure 1. Many HII regions have been observed in the far-infrared where the bulk of the radiation is emitted. There is a fairly tight proportionality between the total infrared emission and the underlying free-free continuum observed at radio wavelengths. Dust temperatures range from 50°K to 250°K with most of the energy emitted by the cold dust. Oddly the most luminous HII region, SgrB2, is also one of the coldest and in this list at 20 μ ranks last.

4. Infrared Emitting Molecular Clouds

Although HII regions are the most luminous galactic sources of infrared emission, there is at least one class of bright extended infrared source which is not related directly to continuum radio sources. These are the Molecular clouds studied by the techniques of radio spectroscopy. The picture that has emerged is one of large dense clouds containing many different molecules and considerable amounts of gas and dust heated by clusters of embedded stars or protostars. Frequently, as in the case of Orion (M42), the molecular cloud or proto-cluster is nearby hotter ionizing stars which have evolved to produce an HII region. The recent discovery of the near-IR lines of H_2 near the K-L Nebulae[1] provides the infrared astronomer a new tool for studying the complex behavior of these exceedingly dense clouds as they evolve to form star clusters and/or HII regions.

5. Diffuse Emission from the Galactic Plane

At wavelengths between 40 and 200 μ the galactic plane appears as a continuous band of emission over more than 30 degrees of longitude. Recent observations of this phenomenon are to be discussed later in this conference by Low, Kurtz, and Poteet, who have measured the peak flux as a function of galactic longitude from 19° through the galactic nucleus to 348°. The continuous band of emission underlying the individual sources is seen clearly. The average width of this band increases from only 0.1 degree to a much larger value as we approach the nucleus. Within ± .5 degrees of the nucleus the width increases dramatically to at least 4 degrees!

6. The Nucleus of our Galaxy

The following points favor the classification of the Galactic Nucleus as unique in the infrared.

1. It contains a cluster of unresolved sources whose properties differ from those of other well observed candidates.

2. The above discrete sources are embedded in (1) a dense assemblage of evolved stars, (2) one or more thermal radio sources, (3) molecular clouds (4) non-thermal radio sources (5) a cloud of dust emitting at about 50°K color temperature.

3. At 100 μ the nucleus appears to have arms which reach out perpendicular to the plane, away from the regions densely populated by stars.

4. Relative to the underlying population of stars, the 10 μ emission is stronger than that observed from M31.

7. Galactic Nuclei

More than 200 extragalactic sources have now been observed at 10 microns. Only the 3 brightest, M82, NGC253, and NGC1068, have been observed at 100 microns.[2] Most of these extragalactic sources are galactic nuclei; they range in 10 micron brightness from less than the brightness of our galactic nucleus to $\sim 10^5$ times. For most galactic nuclei the observed infrared luminosity lies in the range of luminosities expected if stars alone supply the energy. Their spectra and size are consistent with thermal re-radiation models where dust is heated by the stars. The most luminous infrared nuclei appear to exceed the maximum luminosity that can be derived from normal stars. These super-energetic infrared galaxies are thought to be related to the similarly energetic QSO phonomenon.

8. QSO

Only a small number of QSO's have been studied in the infrared because they are so faint. The brightest, OJ287, emitted only 0.7Jy at 10 μ at its peak in 1972. Rieke[3] finds that the active objects have strong apparently non-thermal infrared emission. Thus, as before, this class of objects is easily divided into two or more subclasses even though observations are still relatively few in number.

IV. CONCLUSION

Only 8 distinctly different classes account for all the infrared sources we now know. However, there are many new sources yet to be discovered within these broad classes and it is easy to predict the discovery of sources which do not fit into one of these classes.

Within the galaxy there is at least one new class of infrared nebula still to be detected. Cold clouds of dust heated only by diffuse starlight should soon be detected at 300 microns. It is also possible to postulate entire galaxies embedded in dust so that only their infrared emission is detectable.

As further improvements in instrumentation are made not only will there be many discoveries of new sources but the physics, structure and evolution of these sources will become much better understood and a more complete knowledge of the solid and gaseous molecular constituents of matter in space should follow. Ultimately we can hope to understand the energy sources which power the most energetic systems in the universe.

REFERENCES

1. Gautier, T.N., Fink, U., Treffers, R.R., and Larson, H.P. 1976, in press.

2. Harper, D.A., private communication.

3. Rieke, G.H., private communication.

4. Walker, R.G., and Price, S.D., 1975, AFCRL-TR-75-0373, Hanscom AFB, Mass. 01731.

RESULTS FROM THE AIR FORCE GEOPHYSICS LABORATORY SURVEY CATALOG

Stephan D. Price and Russell G. Walker*

Air Force Geophysics Laboratory
*NASA AMES, Moffett Field

Introduction

Since 1970 the Air Force Geophysics Laboratory has conducted a infrared survey program in order to obtain the spatial and brightness distribution of a representative sample of the types of celestial objects which emit strongly in the 3 to 30 μm spectral region. The survey data, presented herein, were obtained with small cryogenically cooled telescopes flown above the atmosphere by sounding rockets.

Performing the survey above the atmosphere has several important advantages over a ground based experiment. The atmosphere not only absorbs in coming radiation, attenuating the flux and limiting observations to window regions, it also emits radiation. The sky noise caused by the temporal and spatial variations of this emission limit the performance of the ground based system. To minimize this noise, sky cancelling and small fields of view are needed; techniques which strongly descriminate against extended source and are at odds with survey requirements of reasonably large areal scan rates. In space, the atmospheric effects are eliminated and observations can be made in spectral regions in accessable from the ground. Also, the photon background from the telescope can be considerably reduced by cooling to cryogenic temperature. The short time constant and high detectivities exhibited by infrared photoconductors under low background conditions makes practical the use rocket probes to survey the sky.

G. G. Fazio (ed.), Infrared and Submillimeter Astronomy, 13-24.

The Experiment

An important experimental design consideration for this survey was to have the ability to measure the source positions accurately enough to allow ready association with previously catalogued objects and to permit ground based telescopes to acquire the interesting sources for further more detailed, investigation. An accurate geometric reference was established for all the active components of the payload and carefully maintained. The instantaneous field of view was chosen to permit as rapid a scan rate as possible while holding position uncertaintities of a couple of minutes of arc.

The telescope was yoke mounted in a rocket fixed azimuth-elevation system. The payload was spin balanced about the longitudinal or roll axis, which is coincident with the sensor azimuth axis. A fine error guidance sensor, or star tracker, actively held the telescope azimuth axis fixed in celestial coordinates. Telescope was deployed to the desired elevation angle and the payload rotated about the roll axis. The sensor was stepped through an angle slightly less than the total field of view each time the payload rotated 360°. Thus a contiguous sector on the celestial sphere was mapped out.

The star tracker maintained the azimuth axis to a selected "pole" star near the local zenith to within 12 arc seconds. The elevation angle was determined to 30 arc seconds by means of an optical encoder mounted on the deployment shaft. Azimuthal values were obtained to 1 to 1.5 minutes of arc with a stellar aspect sensor by observing stellar transits through an N slit focal plane mask with an S11 phototube. Thus, the geometric line of sight is known to an arc minute accuracy.

Seven successful flights of this experiment were flown out of the White Sands Missle Range, New Mexico during 1971 and 1972. A total of 78 percent of the sky was mapped in three broad band pass colors which had effective wavelengths of 4.2, 11.0, and 19.8 μm. The instantaneous field of view was 3 minutes of arc in the scan direction and 10.5 minutes of arc in the cross scan direction for a solid angle of 4.5×10^{-6} ster. The cross scan resolution was improved by detector to detector overlap to 7.1 and 3.4 arc minutes.

At the conclusion of the northern hemisphere survey, the telescopes were refurbished and modified. The side lobe rejection was improved and the detector widths were increased by 50 percent. Also, the 4.2 μm band was replaced by one which had a 27.4 μm effective wavelength. The telescopes were twice successfully flown from Woomera, Australia. These southern hemisphere flights mapped 36 percent of the celestial sphere, the majority of which lies at negative declinations. The 11.0 μm coverage is now 90 percent of the sky (37000 square degrees).

Data Reduction

The data were digitally filtered to optimize the signal to noise for point sources, then cross correlated with an ideal point source system response. The correlation amplitude gives a best estimate, in the least square sense, for the amplitude of the source. Selection of "possible" sources were based on the signal amplitude to noise ratio and cross correlation coefficient exceeding a value corresponding to a predetermined confidedence level determined under the assumption that the noise was gaussian in distribution. Measurements made on several flights of a source were combined and the signal to noise ratios of the individual detections were added in quadrature if and only if these observations were made in the same color. A second higher confidence level gate was applied and only sources exceeding this value were catalogued.

Gaussian statististics predicted that only a third of the sources exceeding the first gate could be due to noise. The anticipated number of real background sources were too few to account for the difference. Two direct causes which contributed to this excess were identified and treated. One phenomena was due to primary particles, such as cosmic rays, interacting with the detectors. These interactions produced signals characteristic of the system electronic response. This means that pulses due to cosmic rays are significantly narrower and have faster rise times than those due to real sources. It is difficult to distinguish these characteristics for meduim and small amplitude outputs, however, the larger pulses were easily identified and eliminated from the data. Also, inhomogenieties in emission from the earth and earth's limb detected by the side lobe response of the telescope produced spurious pulses. Since these signals occur when the noise is high and are wide angle effects and measured as extended they are also easily recognized. Approximately 5 to 10 percent of the "possible" sources were estimated to be due to cosmic rays and another 10 to 20 percent could be due to the granularity in the earth's emission.

The gating at two different confidence levels was done to allow the fainter sources to be selected by rescanning and to account for a source not having the same signal to noise ratio each time it was measured. This variability is due to the source being intrinsically variable, the responsivily varying from detector to detector and, in some cases, flight to flight, and/or the non-stationary behavior of the noise due to earth radiation detected through the telescope side lobes. In order to account for the latter two effects the signal to noise ratio was calculated for those detectors on each flight which scanned a previously detected source position but did not redetect it. If the calculated signal to noise did not exceed the second ratio, the noise was considered too high for conformation. These sources which could have been seen but were not were considered unconfirmed and rejected from the catalog. This was an important criteria in eliminating the spurious sources discussed above.

Results

The revised and extended AFGL Infrared Sky Survey catalog of Price and Walker (1976) contains measurements on 2361 sources detected in one or more of the survey bands. The total area, multiply scanned area and the number of sources observed in each color are shown in table 1. Approximately two-thirds of the area scanned in each color was surveyed at least twice. The catalog is statistically complete down to limiting magnitudes of $M(4.2) \simeq +1.3$, $M(11.0) \simeq -1.2$ and $M(19.8) \simeq -3.0$. These limits are defined as the flux below which the observed log N-log H relation departs from linearity. The faintest sources in the catalog are three to four times lower than these limits. The catalog is incomplete at the faintor levels due to detector to detector differences in responsivity and non-stationary behavior of the noise during a flight.

Table I

Area of sky mapped for each wavelength band

Percentage of sky Color/Minumum time scanned	1	2	3	4	Total No. of Sources
4 μm	78	50	20	6.5	1981
11.0 μm	90	60	34	11.4	1149
19.8 μm	87	50	28	8.7	645
27.4 μm	37	4	--	----	72

It is emphasized that in a survey that does not regularly monitor all sections of the sky, variable sources will be preferentially detected at their brightest. Variations of up to 1.5 magnitudes at 11 μm are known to exist for sources such as IRC + 10011. Also, Merrill (1975) measured several of the unidentified sources in the AFCRL catalog (Walker and Price, 1976) to vary by as much as a factor of two at 11 μm with time scales of a year.

The measured brightness of extended sources is, of course, a strong function of the instantaneous field of view. However, the sensor system can only distinguish as extended those sources which are extended two to three minutes of arc in the scanning direction. Smaller sources would not be seen as extended. Recently, Sayre

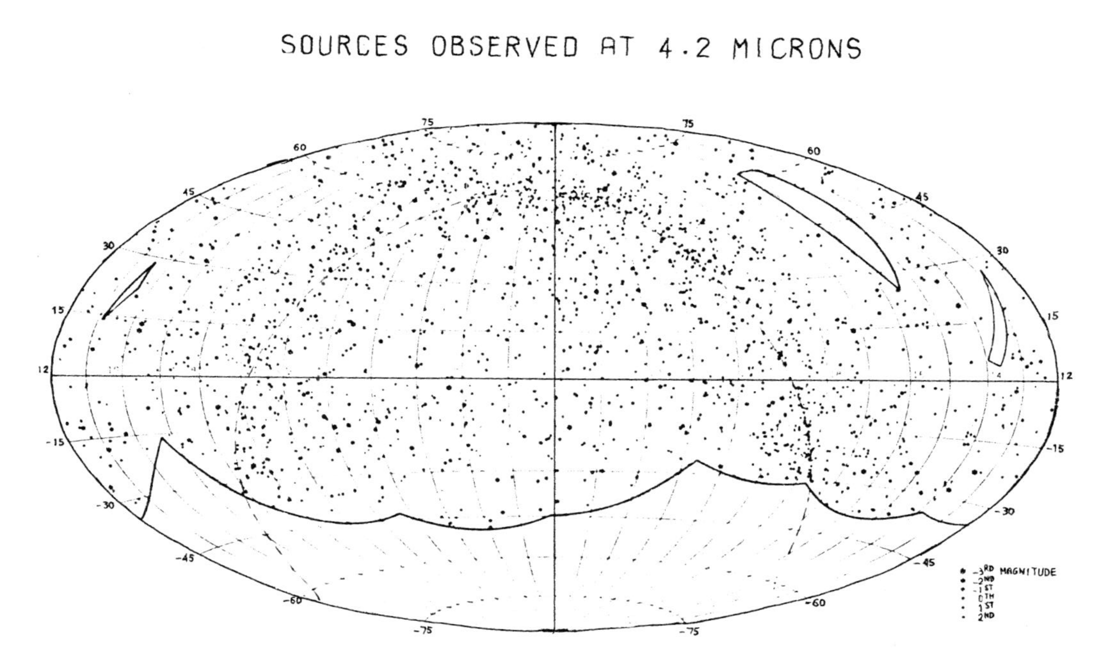

Figure 1 - Distribution of sources observed at 4.2 μm on an Aitoff equal area all sky projection. Heavy lines are the survey limits, dashed line is the galactic equator.

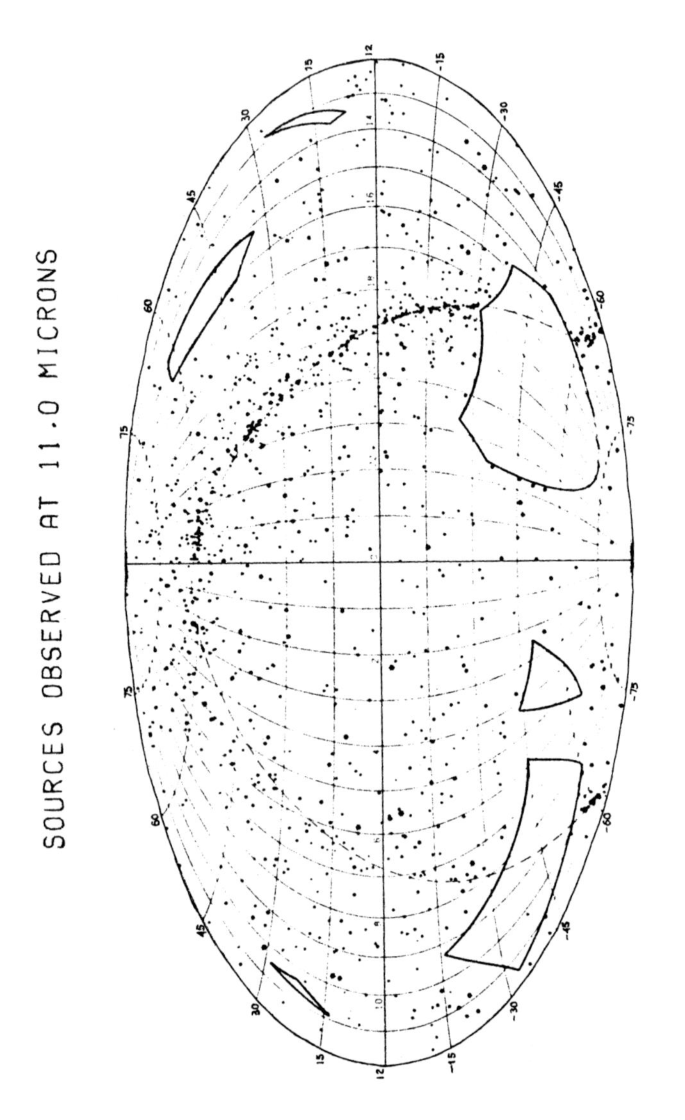

Figure 2 - Distribution of 11.0 μm sources.

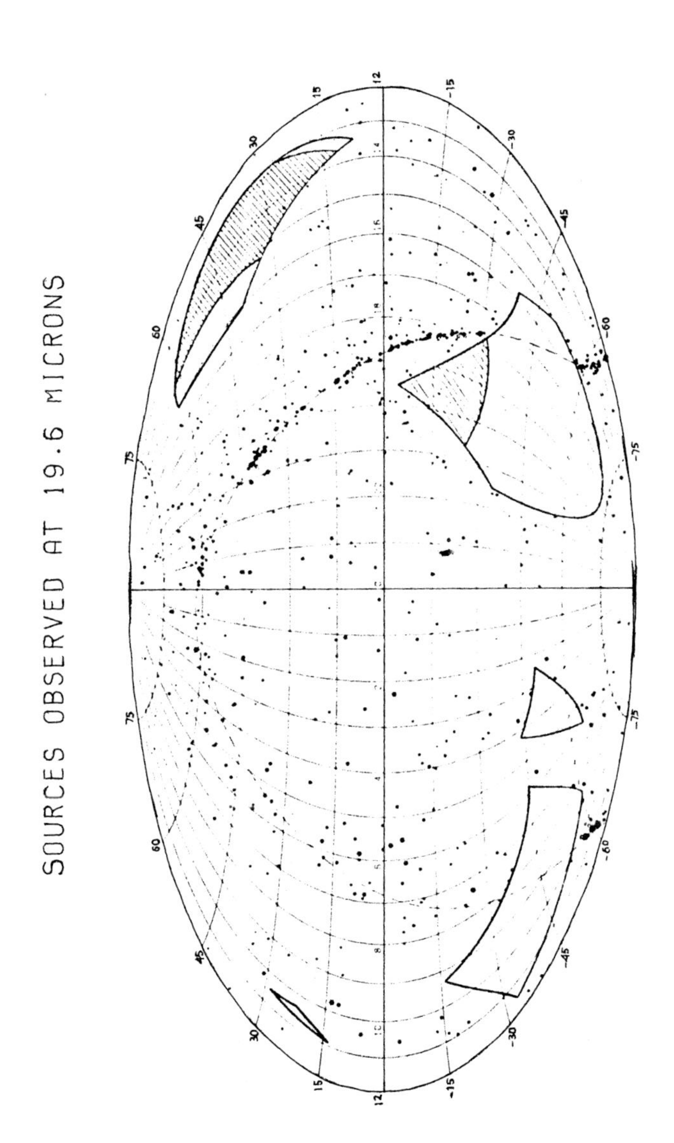

Figure 3 - Distribution of 19.8 μm sources. The dashed area was not completely covered due to a failure of several of the 19.8 μm channels on the flight that covered these regions.

et al (1976), has found that the detector response may be higher to extended sources than for point sources. The sources observed at 4.2, 11.0, and 19.8 μm are plotted on Aitoff equal area all sky projections in figures 1, 2, and 3 respectively. The concentration of sources to the galactic plane evidenced in these plots is more markedly depicted in figures 4, 5, 6, and 7 which are histograms of the sources in each color as a function of galactic latitude. The upper values are for the total number of sources in each latitude band, while the cross hatched areas are for sources which are not associated with IRC objects (Neugebauer and Leighton, 1969; Neugebauer, 1973) or a Bright Star (Hoffleit, 1964). Sixteen percent of the 4.2 μm, 32 percent of the 11.0, 52 percent of the 19.8 μm and 75 percent of the 27.4 μm sources are thus unassociated indicating that the percentage of non-stellar objects increases with increasing wavelength. Also, a source associated with the IRC or Bright Star catalog will usually have a 4.2 μm observation; over 98 percent of these sources have the short wavelength measurements. It is interesting to note that only 25 of the 95 sources in the Walker and Price catalog which are unassociated and confirmed from the ground do not possess a short wavelength detection. Thus, the ground based confirmation seem to do better with star like sources.

The galactic latitude distribution of the 104 sources that remain in the present catalog out of the 327 in the Walker and Price catalog which were not confirmed from the ground by various groups (Keinniann, 1975; Low, 1973; Hackwell, 1976; Merrill, 1976) is shown in figure 8. The sources are highly concentrated to the plane, 60 percent be within 5 degrees and 80 percent within 20 degrees of the galactic plane. A comparison of this distribution with the one in figure 5 shows that the distribution of these unconfirmed sources in not different than that for all the sources. Over half the unconfirmed sources are either measured as extended, associated with HII regions or were observed three times in a common color. The cross hatched area dipicts, for comparison, the distribution with these sources removed.

The data contained in the present catalog was optimized for point source detection. The electronic system response was designed to as low a frequency as possible to preserve information on widely extended sources. Diffuse emission from the galactic plane and HII region such as the Orion Nebula have been mapped at 11 and 19.8 μm but the data still has to be processed to optimize these observations.

Acknowledgment: We would like to express our appreciation to Robert Pelzmann for all the graphics in this report.

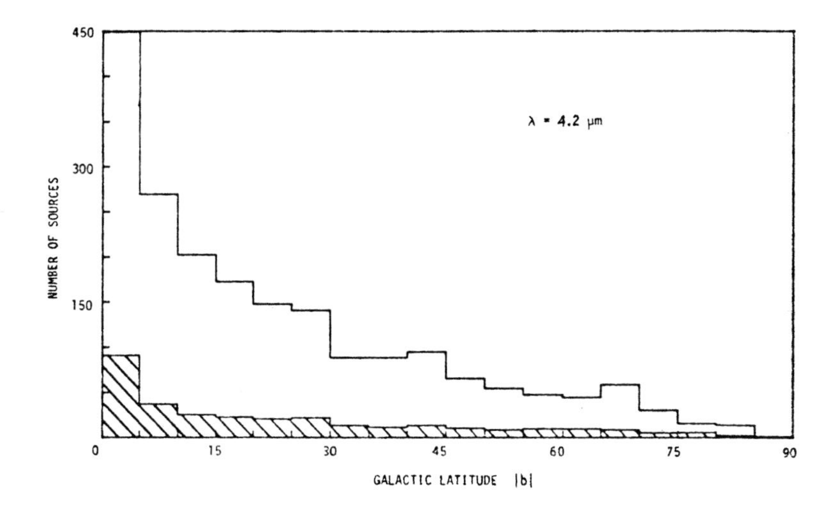

Figure 4 - Histogram of 4.2 μm sources as a function of the absolute galactic latitude. The upper area represents all the sources, the lower dashed area represents only the "non-identified" sources.

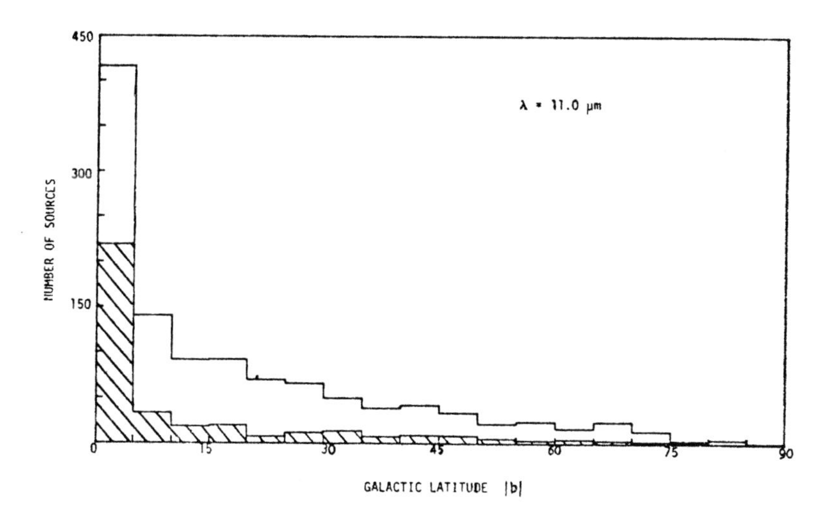

Figure 5 - Same as figure 4 but for the 11 μm source.

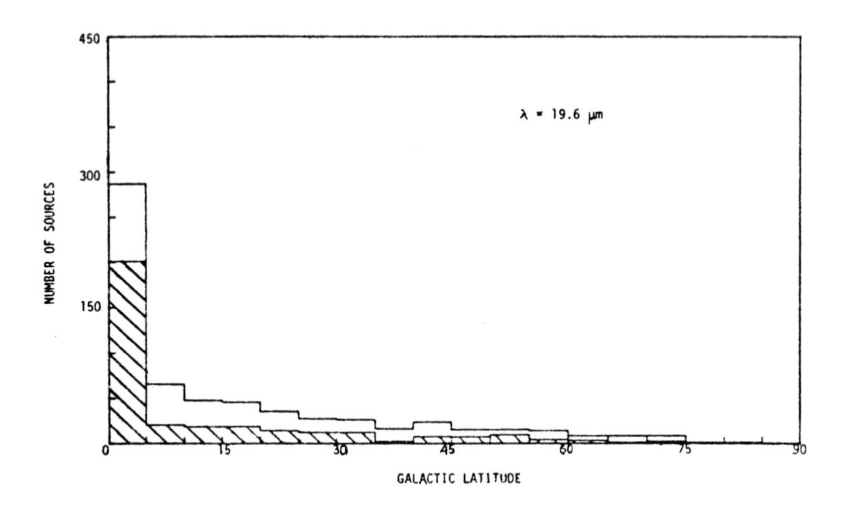

Figure 6 - Same as Figure 4 but for the 19.6 μm sources.

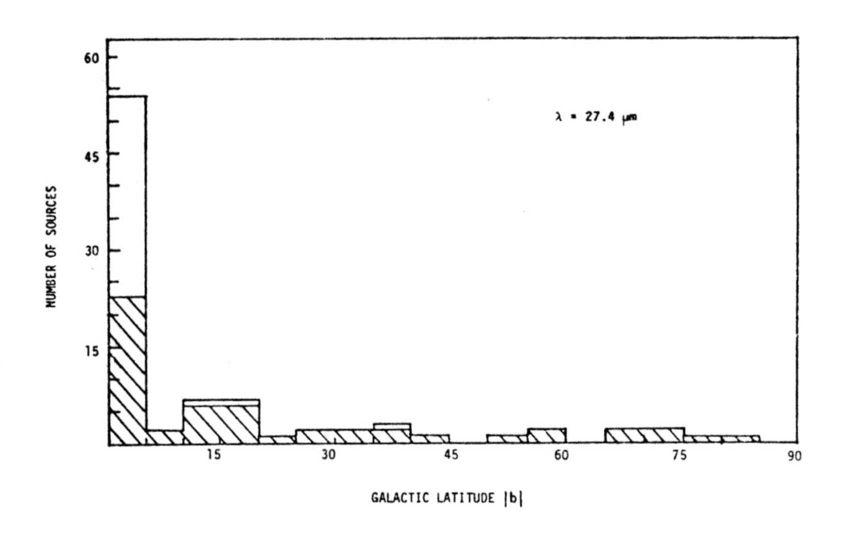

Figure 7 - Same as figure 4 but for the 27.4 μm source.

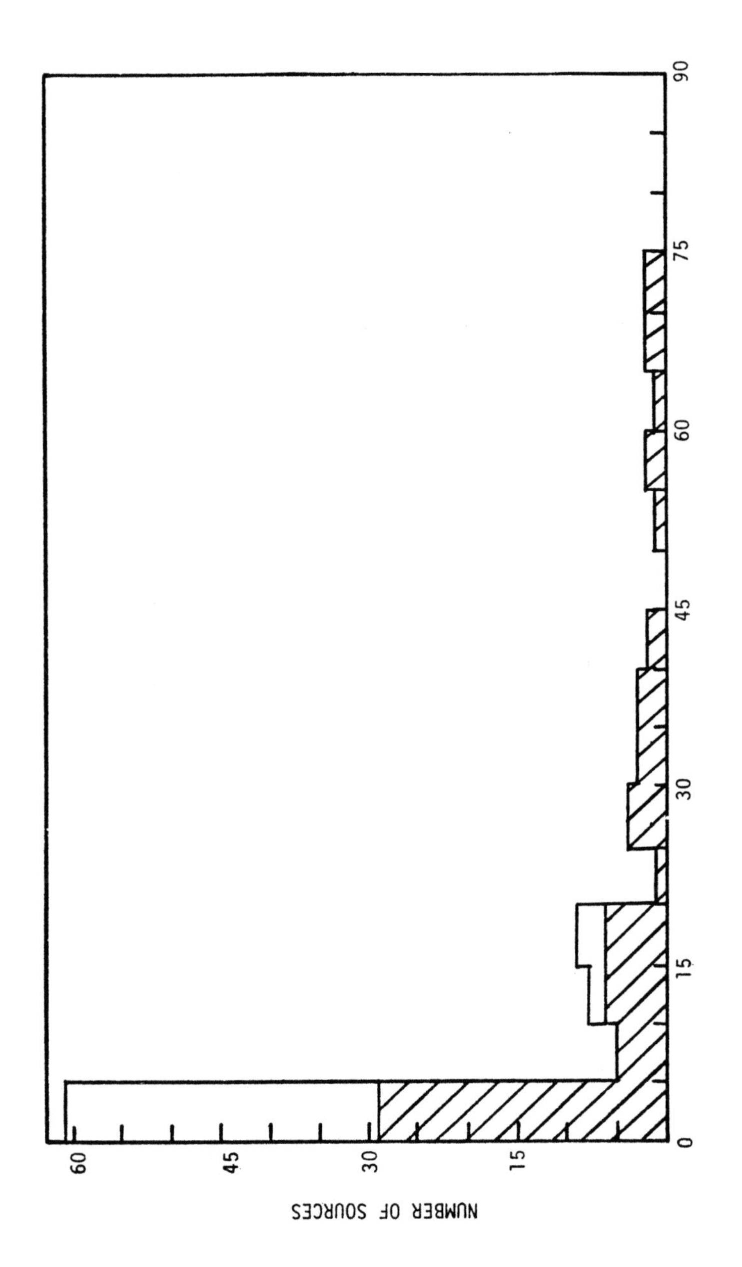

Figure 8 - Histogram of the sources in the catalog searched for but not found from the ground as a function of the absolute value of the galactic latitude.

References

Hackwell, J.A., 1976, private communication.
Hoffleit, D., 1964, Catalogue of Bright Stars, 3rd ed.
Kleinmann, S.G., 1975, private communication.
Low, F.J., 1973, AFCRL-TR-73-0371.
Merrill, M., 1976, private communication.
Neugebauer, G., 1973, private communication.
Neugebauer, G., and Leighton, R.B., 1969, NASA, SP-0347
Sayre, C., Arrington, D., Eisenman, W., and Merriam, J., 1976, preprint from March 1976 IRIS meeting on Detector Effects.
Walker, R.G., and Price, S.D., 1975, AFCRL-TR-73-0373.

FURTHER OBSERVATIONS OF NEW SOURCES IN THE AFCRL SURVEY

S. G. Kleinmann
Physics Department and Center for Space Research
Massachusetts Institute of Technology
Cambridge, Massachusetts 02139

In this paper I will summarize at least some of the highlights of the follow-up observations that have been obtained for the new sources discovered in the AFCRL rocket infrared Sky Survey (Walker and Price, 1975). These observations are designed to determine the natures of the new AFCRL sources, in expectation that some of the sources in the Catalogue may represent new classes of astrophysical objects. To do this, observations covering virtually the entire gamut of available astronomical tools are being employed. I hope to show that the discovery of a number of truly remarkable objects in the AFCRL Survey amply justifies further efforts to explore it.

The Ground-Based Searches

Before I discuss these observations, it is useful to describe how certain objects from the AFCRL Catalogue have been selected for study. Most of the follow-up observations utilize optical systems with beams that are narrow compared to the 3'x10' system used in the AFCRL Survey. As a result, the first task in studying new AFCRL sources is to obtain more accurate locations for the sources than those given in the Catalogue. The most commonly used means of obtaining improved positions for the AFCRL sources involves raster scanning AFCRL error boxes with infrared photometric systems usually designed to detect sources $\gtrsim$ 1 mag. fainter than the limits of the AFCRL Survey. In this way, more than 85 new sources have been located with accuracies ranging from ± 1" to ± 30".

An additional outcome of the efforts to improve the error boxes of AFCRL sources has been to show that many--in fact, the

G. G. Fazio (ed.), Infrared and Submillimeter Astronomy, 25-33.

majority--of the ∿400 sources, for which searches have been made, cannot be found by ground-based searches (Lebofsky, Kleinmann and Rieke, 1975; Lebofsky et al. 1976; Gehrz and Hackwell, 1976). Table I summarizes these results.

Table 1: Summary of Searches for New AFCRL Sources

11μ Sources		
Total number of new sources	=	657
Total number of confirmed sources	=	87
Total number of unconfirmed sources	=	281
4μ Sources		
Total number of new sources	=	799
Total number of confirmed sources	=	3
Total number of unconfirmed sources	=	27

As shown in Figure 1, the confirmed sources lie concentrated near the galactic plane, while the unconfirmed sources are more randomly distributed. This fact was first noted by Low (1973) in his study of sources in a preliminary version of the AFCRL Catalogue.

The cause of these troubling facts have been investigated by M. Hauser, F. Low, G. Rieke and S. Price in a review of the

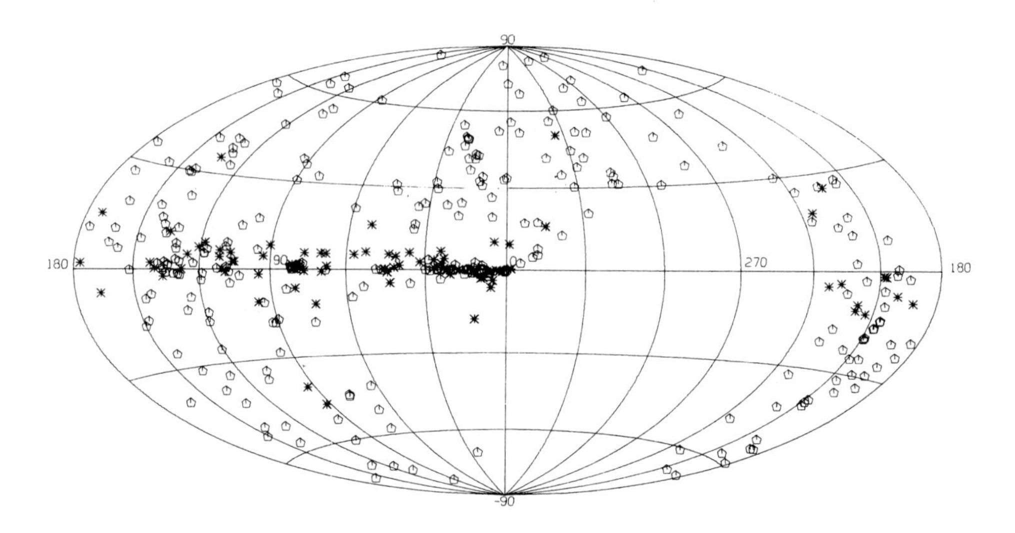

Figure 1: An Aitoff projection in galactic coordinates of the new 11μ AFCRL sources for which ground-based searches have been made. Pentagons represent unconfirmed sources, asterisks represent confirmed sources.

classified raw data and reduction techniques of the AFCRL Sky Survey. They found that many of the sources reported in the Catalogue had been detected with low confidence. This result was not entirely new or surprising since the AFCRL data reduction included a deliberate effort to avoid over-estimation of the threshold for detection. In this way, faint, interesting sources would be included in the Catalogue, only at the expense of including a number of spurious sources as well. It will be shown later in this paper how this philosophy has been a fruitful one. In the meantime, the NASA report prepared by Hauser _et al_. suggests that the number of spurious sources in the Catalogue is likely to be sufficiently large to account (almost) entirely for the poor success attained by observers making searches for new AFCRL sources. In addition, Hauser _et al_. reported that the discrepant galactic distributions of the confirmed and unconfirmed sources could be understood in terms of the slightly different survey system characteristics of each of the rocket flights.

Class Properties of the New AFCRL Sources

On this basis, it seems plausible that the 87 sources confirmed so far will represent at least a major fraction of all the new (11μ) sources that will be found in the AFCRL Catalogue. Therefore, it is not unreasonable to suppose that the class properties of the new AFCRL sources may already be emerging from observations of these 87. These are summarized below.

1. Galactic Distribution

As I noted earlier, the confirmed sources are generally located near the galactic plane. If we infer that these sources are located in spiral arms, then the AFCRL survey must penetrate quite deeply into the Milky Way. Therefore, many of the new sources must be luminous objects ($>10^3$ $L_\odot$), since they are generally brighter than 100 Jy at 11μ. Most of these luminous objects are probably evolved massive stars, since only a few of them are located near HII regions, or other regions of recent star formation. (It should be noted that these generalities are subject to the criticism that the AFCRL Catalogue may be incomplete above the galactic plane, at the average flux level of the confirmed sources.)

2. Spectra

In the infrared, the new sources exhibit continuum radiation that can be characterized by color temperatures $100<T(°K)<800$. Low _et al_. (1976) and Lebofsky _et al_. (1976) found broad 10μ absorption or emission features superposed on the continua of most of the 61 AFCRL sources they found by means of narrow-band photometry. Higher-resolution, mid-infrared spectroscopy has been carried out for ∿20 sources so far, both at the USCD/U. Minn. telescope (Merrill and Soifer, 1974; Cohen *et al.* 1975) and at KPNO (Kleinmann *et al.* 1976). These data confirm the common presence of 10μ dust signatures in the spectra of AFCRL sources, and enable us to distinguish carbon stars, showing SiC features, from oxygen-rich stars. However, a number of the coldest sources appear to have continua without strong features in the 10μ region--a fact that may play a significant role in future studies to determine how to classify stars found in infrared sky surveys.

In the near infrared, many of the AFCRL sources exhibit strong CO and H_2O absorption, which are indicative of the luminosity, temperature, and composition of the stars. Ice (3.1μ) absorption is seen in the spectra of a few stars--generally those found near, or associated with, regions of active star formation.

Optically, about half of the AFCRL sources have been tentatively identified with faint, red stars (Kleinmann and Lebofsky, 1975; Lebofsky and Kleinmann, 1976). Most of the others lie in obscured areas or lie in particularly crowded areas, where high positional accuracy is needed to make plausible identifications; this fact suggests that identifications from the Palomar plates or near infrared plates will eventually be available for the majority of new AFCRL sources, so that a survey of their general optical properties is feasible. Cohen and Kuhi (1976) obtained optical spectrophotometry of 13 objects identified with CRL sources; because of reddening or confusion, about half of these objects may be incorrect identifications, and Cohen and Kuhi derived relatively early (K7-M2) spectral types for them. The others are very late M giants or carbon stars. One is a late WC-type star. Again, these data imply that many of the new CRL sources are cool, luminous stars.

In contrast to these red stars, reflection nebulae have been found near several AFCRL sources (Cohen *et al.* 1975; Kleinmann and Lebofsky, 1975; Ney *et al.* 1975; Westbrook *et al.* 1975), and high resolution spectra of a few of these indicate that the stars illuminating them have spectral types ranging from B to late F.

So far, extensive spectroscopic studies at microwave frequencies have been obtained for only a few of the new AFCRL sources. However, P. Myers, S. Zisk, A. Hashick, D. Sargent and I are carrying out surveys of all the new AFCRL sources for OH and H_2O emission and these are partially complete. So far, 14 new OH masers have been found, out of 86 confirmed new sources. These detections significantly increase the number of known OH/IR stars. The sources that are associated with masers are among the brightest objects found by the AFCRL survey and may be significantly colder than previously known OH/IR stars. The large, average velocity separation of their two OH emission components provides more evidence that these sources are luminous stars (Dickinson and

Chaisson, 1973) with high mass loss rates. About half of the confirmed sources have been searched for H_2O emission, with the result that 8 new H_2O masers were found. Some of these are highly variable, implying that continued searches for H_2O maser emission, especially from stars already found to be OH masers, are warranted.

3. Temporal Behavior

Of the 19 sources which Low et al. (1976) observed on two different occasions, a third varied by a factor of two or more. The time difference between the two sets of observations was one year. This behavior is reminiscent of the variability observed by Strecker and Ney (1974) among many of the new stars that were found in the Two Micron Sky Survey.

Taken together, these data suggest that most of the new AFCRL sources are late-type giants and super-giants similar to the sources of high I-K index discovered in the Two Micron Sky Survey (Neugebauer and Leighton, 1969).

Unusual Objects

The question still remains, has the AFCRL Sky Survey resulted in the discovery of any new kinds of astrophysical objects? The following descriptions of the few extensively studied AFCRL sources provides the only answer.

Several unusual objects that are possible protostars were discovered by the AFCRL survey, and CRL 877 stands out among these. This source is located near the center of a cluster of stars associated with reflection nebulae, called the Mon R2 association (Racine, 1968). Because of this, CRL 877 was one of the first new AFCRL sources for which we searched. It was found to be a bright object with strong 3.1μ ice band absorption, an apparent characteristic of the cold, dense clouds in which young, infrared sources are found (Merrill, Russell and Soifer, 1975). This cloud was detected as a strong CO source by R. Loren during his survey of R-associations, and by Kutner and Tucker (1974) in their program to map the Orion region in CO. On the basis of these microwave studies, Beckwith et al. (1976) mapped the Mon R2 region in the infrared and found a small cluster of cold, infrared sources. Recently, Low and collaborators (private communication) found a bright, far-infrared source associated with Mon R2. These discoveries suggest that the generic relationship between CRL 877 and the Kleinmann-Low nubula in Orion is a close one. The existence of H_2O masers in the Mon R2 cloud leads to the argument that at least some of the sources in it are young, perhaps $\sim 10^3$ years (Burke, 1973). In contrast, the visible stars in the R-association must be near 10^7 years old, evidence that star formation in this region was triggered at disparate epochs. The significance of the discovery of Mon R2 is that observations of objects which are members of a single cluster or association, but which represent a wide variety of evolutionary progress, are the basic data upon which theories of early stellar evolution must rely, and in this respect Mon R2 must represent one of the richest clusters yet observed.

Certainly the Egg Nebula (CRL 2688) is also widely recognized as one of the most intriguing of the new sources found in the AFCRL Sky Survey. This object has been studied photometrically by Ney et al. (1975) who found that it must consist of a late-type star surrounded by a thick toroidally-shaped dust cloud. The infrared emitting region of the toroid is extended with $\theta \sim 2''$. A large tenuous cloud surrounds the object and is seen as a reflection nebula illuminated by the starlight that escapes above and below the ring of dust. The discovery of this object has provided dramatic evidence supporting earlier hypotheses that circumstellar envelopes of infrared stars must be flattened. The large optical depth of the toroid produces a featureless, mid-infrared spectrum (Forrest et al. 1976) but the chemical nature of the cloud has been deduced from optical spectroscopy of the reflection nebula (Crampton, Cowley and Humphreys, 1975) and by detection of a molecular cloud association with the source (Lo and Bechis, 1976 and Zuckermann et al. 1976). These observations show that the $0.1\ M_{\odot}$ cloud is carbon-rich, and, in fact has led to the suggestion that the source may be the progenitor of a planetary nebula.

The discovery by Westbrook et al. (1975) of an object, CRL 618, which is similar to CRL 2688 in its geometry and lack of infrared spectral features, but similar to a planetary nebula in its optical spectrum, tends to support that hypothesis. However, other objects with similar geometries but different spectral properties have been found, and their evolutionary status is not clear. In particular, M1-92, which exhibits a double-reflection nebula of early spectral type, surrounds a star showing an expansion velocity near 500 km/sec (Herbig, 1975). Thus, dynamically, M1-92 is clearly distinguished from planetary nebulae, which typically have expansion velocities $\sim$20 km/sec. The discovery of an unusual Type I OH maser associated with this source (Lepine and Rieu, 1974) casts further doubt on the possibility that it is related to planetary nebulae.

Another object associated with multiple reflection nebulae is CRL 437. The reflection nebula was found by Kleinmann and Lebofsky (1975) in their search for optical counterparts for new AFCRL sources by means of near-infrared photography with image tube cameras at KPNO. Direct observation of the brightest part of the complex nebula on the KPNO 1.3-m indicated at first, that the source was ten times fainter than reported in the AFCRL Catalogue. This cast doubt on the identification of the reflection nebula with the CRL source, especially since this nebula was 3.6' west of the CRL position. However, no other object brighter than the source near the reflection nebula was found in a 9'x9' box centered on the CRL position. It should be noted that all these observations were made using a bolometer system with a 10μ (N-band) filter, a 12" beam and 20" throw. R. R. Joyce and F. C. Gillett then showed that CRL 437 is extended at 2.2, 3.5 and 10μ with $\theta \sim 20''$, which explained the discrepancy between our first flux measurement and that given in the AFCRL Catalogue. This discovery, the details of which will be reported at the 148th AAS Meeting (Kleinmann _et al_. 1976), is especially surprising since neither Br α nor radio continuum emission could be detected from this source. Thus, unless CRL 437 consists of a compact cluster of stars there is no obvious mechanism by which the dust is heated! This extended infrared emission region is surrounded by an even more extended (6'x6') molecular cloud, discovered by Barrett _et al_. (1976). Interpretation of the profiles of the microwave emission lines indicates that the cloud is undergoing slow expansion and rotation, so the source(s) at its center may be evolved. Appropriate model-building for this obviously peculiar object awaits further data, including a determination of its total luminosity by far-infrared observations, mid-infrared maps with high spatial resolution and spectral resolution, and optical spectroscopy and polarimetry of each of the nebulae near CRL 437.

For the present, however, the information content of these observations of CRL 437 lies in the fact that it is a dramatic example of the high probability of serendipic discoveries. This is one of the new AFCRL objects that would _not_ have been found in any of the on-going programs to confirm CRL sources because it is extended, faint, and relatively distant from the AFCRL position; it might easily have been overlooked since it was detected on only one out of three rocket flights in the AFCRL Survey and is reported at a level nearly 1 mag. below the statistical limit of the Catalogue. Thus, the discovery of this source is a direct consequence of the decision by Walker and Price to include objects in the Catalogue that were detected below the statistical limits of the survey. These facts suggest that, although many of the new AFCRL sources are probably spurious, additional searches for these objects may prove to be scientifically profitable. These searches should utilize a variety of techniques, and should be aimed at studying the new 4μ and 20μ sources in the Catalogue, as well as the intensely studied 11μ sources.

Acknowledgement

This paper benefitted from discussions with M. Hauser and F. Low. This work was supported by the National Aeronautics and Space Administration under Grant NSG-7186.

References

Barrett, A.H., Ho, P.T.P., Martin, R.N., and Schneps, M.H. 1976, to be presented at 148th A.A.S. Meeting.

Beckwith, S., Evans II, N.J., Becklin, E.E., and Neugebauer, G. 1976, preprint.

Burke, B.F. 1973, in I.A.U. Symposium No. 60, ed. F.J. Kerr and S.C. Simonson III (D. Reidel Publishing Co., Dordrecht, Holland), p. 267.

Cohen, M., Anderson, C., Cowley, A., Coyne, G., Fawley, W.M., Gull, T., Harlan, E.A., Herbig, G.H., Holden, F., Hudson, H.S., Jakoubek, R.O., Johnson, H.M., Merrill, K.M., Schiffer, F.H., Soifer, B.T., and Zuckermann, B. 1975, Ap. J., 196, 179.

Cohen, M.H. and Kuhi, L.V. 1976, Pub. A.S.P., in press.

Crampton, D., Cowley, A.P., and Humphreys, R.M. 1975, Ap. J. (Letters), 198, L135.

Dickinson, D.F. and Chaisson, E.J. 1973, Ap. J. (Letters), 181, L135.

Gehrz, R.D. and Hackwell, J.A. 1976, Ap. J. (Letters), in press and private communication.

Herbig, G.H. 1975, Ap. J., 200, 1.

Kleinmann, S.G. and Lebofsky, M.J. 1975, Bull. A.A.S., 1, 433.

Kleinmann, S.G. and Lebofsky, M.J. 1975, Ap. J. (Letters), 201, L91.

Kleinmann, S.G., Capps, R.W., Gillett, F.C., Grasdalen, G.L., Joyce, R.R., Pipher, J.L., and Sargent, D.G. 1976, to be presented at 148th Meeting of A.A.S.

Kutner, M.L. and Tucker, K.D. 1974, Bull. A.A.S., 6, 341.

Lebofsky, M.J. and Kleinmann, S.G. 1976, Ap. J., in press.

Lebofsky, M.J., Kleinmann, S.G., Rieke, G.H., and Low, F.J. 1976, Ap. J. (Letters), in press.

Lepine, J.R.D. and Rieu, N.Q. 1974, Astr. and Ap., 36, 469.

Lo, K.Y. and Bechis, K.P. 1976, Ap. J. (Letters), 205, L21.

Low, F.J. 1973, AFCRL-TR-0371.

Low, F.J., Kurtz, R.F., Vrba, F.J., and Rieke, G.H. 1976, Ap. J. (Letters), in press.

Merrill, K.M. 1975, Bull. A.A.S., 1, 433.

Merrill, K.M. and Soifer, B.T. 1974, Ap. J. (Letters), 189, L27.

Merrill, K.M., Russell, R.W., and Soifer, B.T. 1975, Bull. A.A.S., 7, 429.

Racine, R. 1968, A. J., 73, 588.

Neugebauer, G. and Leighton, R.B. 1969, NASA SP-3047.

Ney, E.P., Merrill, K.M., Becklin, E.E., Neugebauer, G., and Wynn-Williams, C.G. 1975, Ap. J. (Letters), 198, L129.

Strecker, D.W. and Ney, E.P. 1974, A. J., 79, 797.

Walker, R.G. and Price, S.D. 1975, AFCRL-TR-75-0373.

Westbrook, W.E., Becklin, E.E., Merrill, K.M., Neugebauer, G., Schmidt, M., Willner, S.P., and Wynn-Williams, C.G. 1975, Ap. J., 202, 407.

Zuckermann, B., Turner, B.E., Gilra, D., Morris, M., and Palmer, D. 1976, Ap. J. (Letters), 205, L15.

LONG-TERM INFRARED MONITORING OF STELLAR SOURCES FROM EARTH ORBIT

S.P. Moran*, T.F. Heinsheimer**, T.L. Stocker**,
S.P.S. Anand***, R.D. Chapman***, R.W. Hobbs***,
A.G. Michalitsanos***, F.H. Wright+, and S.L. Kipp++

ABSTRACT

The 2.7μm radiation emitted by a selection of late-type variable stars during three years was monitored with U.S. Air Force satellite instrumentation. The data comprise the most detailed and extensive collection of infrared light curves reported for such stars and represent the first systematic infrared astronomy performed from Earth orbit. A well-defined linear increase in flux density is found to characterize the first three tenths of each cycle following infrared minimum of a long-period variable star. Examination of the data on long-period and semi-regular stars also shows marked differences between successive whole cycles, although there are certain phases at which the flux density repeats rather precisely. Large convective cells, as predicted for red supergiants, may couple with stellar pulsation in a manner that accounts for this phenomenon. The stars observed include known sources of circumstellar microwave line emission that may be pumped by the variable infrared continuum near the wavelength of observation. Strong coupling of large-scale convection and pulsation can give rise to the circumstellar clouds and, under certain conditions, may even provide the mechanism for ejection of material to form a planetary nebula.

*NASA-Goddard Space Flight Center, Greenbelt, Marylamd, USA and University of California, Los Angeles, California, USA.
**The Aerospace Corporation, El Segundo, California, USA.
***NASA-Goddard Space Flight Center, Greenbelt, Maryland, USA.
+Aerojet ElectroSystems Company, Azusa, California, USA.
++Wesleyan University, Middletown, Connecticut, USA.

G. G. Fazio (ed.), Infrared and Submillimeter Astronomy, 35.

BALLOON OBSERVATION OF THE MILKY WAY AT WAVELENGTH 2.4 μm

S. Hayakawa, K. Ito, T. Matsumoto, and K. Uyama

Department of Physics, Nagoya University, Nagoya, Japan

ABSTRACT

Raster scanning of the Milky Way in the galactic longitude range 23° - 75° was performed at a wavelength 2.4 μm by a balloon-borne telescope cooled by liquid nitrogen. With the field of view $3^{o} \times 3^{o}$ the galactic plane was found as narrow as 6° (FWHM). The flux at the galactic equator was observed to be 6×10^{-10} W cm^{-2} μm^{-1} at $\ell = 23^{o}$ and to decrease rather steeply to 8×10^{-11} W cm^{-2} m^{-1} at $\ell = 75^{o}$. The longitude dependence is similar to that of the brightness temperature of radio continuum and shows humps at the same positions as was found for the intensity of 2.6 mm line of CO. This suggests the contribution of young objects to the infrared brightness. The isophotes of the infrared intensity are compared with that expected for models of the galaxy.

G. G. Fazio (ed.), Infrared and Submillimeter Astronomy, 36.

PREDICTION OF THE DIFFUSE FAR-INFRARED FLUX FROM THE GALACTIC PLANE

F. W. Stecker & G. G. Fazio

NASA Goddard Space Flight Center, Greenbelt, Md. (USA)
Center for Astrophysics, Cambridge, Ma. (USA)

ABSTRACT

A basic model and simple numerical relations useful for future far-infrared studies of the galaxy are presented. Making use of recent CO and other galactic surveys, we then predict the diffuse far-infrared flux distribution from the galactic plane as a function of galactic longitude ℓ for $4^o \leq \ell \leq 90^o$ and the far-infrared emissivity as a function of galactocentric distance. Future measurements of the galactic far-infrared flux would yield valuable information on the physical properties and distribution of dust and molecular clouds in the galaxy, particularly the inner region.

G. G. Fazio (ed.), Infrared and Submillimeter Astronomy, 37. *All Rights Reserved.*

FAR INFRARED EMISSION OF MOLECULAR CLOUDS

C. E. Ryter and J. L. Puget

Centre d'Etudes Nucleaires de Saclay, France
Observatoire de Paris-Meudon, France

ABSTRACT

Data presently available on far infrared (10 - 300 μm) thermal emission and carbon monoxide millimeter radiation are compiled in order to generate a sample of objects where the thermal radiation of the dust mixed with the molecular hydrogen can be quantitatively studied. A list of 10 massive molecular clouds is obtained. A luminosity, L^H_{IR}, normalized per hydrogen atom, is introduced and found to range from 0.5 to 5 x 10^{-30} W (H-atom)$^{-1}$, confirmed by a value L^H_{IR} = 2.4 x 10^{-30} W (H-atom^{-1}), which can be deduced from a measurement of the diffuse galactic far infrared emission (Pipher 1973).

The quantity L^H_{IR} is more than one order of magnitude above the value expected from heating of the dust by the average starlight density, u $\approx$ 0.5 eV cm^{-3}. The clouds are clearly heated from the interior, by newly born stars still unobservable. But any attempt to infer the power released in the clouds using published star formation rates deduced from star count in the solar neighborhood falls short by a factor $\sim$ 10 or larger.

Then it was assumed that the dust is made of a mixture of "silicate" and ice grains matching the abundances deduced from spectral (3.1 and 10 μm) observations; however, the "silicate" is supposed to be in an amorphous state (recently observed), in which the far infrared emissivity is substantially enhanced. The temperature deduced from the observed value of L^H_{IR} and the grain model is found to be in very good agreement with the temperatures actually obtained from multicolor photometry.

G. G. Fazio (ed.), Infrared and Submillimeter Astronomy, 38.

HIGH-ENERGY GAMMA-RAY RESULTS FROM THE COS-B SATELLITE AND INFRARED ASTRONOMY

Jacques Paul

Service d'Electronique Physique Centre d'Etudes
Nucleaires de Saclay
Gif/Yvette, France

ABSTRACT

The European Space Agency satellite COS-B carries a single experiment devoted to the observation of celestial high-energy (> 30 MeV) gamma-rays. Some of the preliminary results indicate that a fraction of the galactic gamma-rays originate in close-by regions (distance $\leq$ 1 kpc). The interpretation of these results in correlation with infrared observations will permit a study of the content structure of the local interstellar medium.

G. G. Fazio (ed.), Infrared and Submillimeter Astronomy, 39.

SEARCH OF "INVISIBLE" SUPERNOVAE

O. F. Prilutskii

Academy of Sciences of the USSR
Space Research Institute
Moscow, USSR

ABSTRACT

The possibility of observing the outburst of Supernovae at the remote Galaxy regions, invisible within optical band because of interstellar extinction using thermal IR-radiation of interstellar dust heated by Supernova outburst is discussed in this paper. The investigation of this phenomenon by means of cooled IR-telescopes will allow the Supernova outburst parameters to be determined and the characteristics of the interstellar dust distribution to be studied. The similar effects can be detected in the vicinities of the well-known remnants of Supernovae, in particular, near Tycho Brahe and Cassiopeia A Supernovae.

G. G. Fazio (ed.), Infrared and Submillimeter Astronomy, 40.

PART II

G A L A C T I C   S O U R C E S

INFRARED EMISSION FROM HII REGIONS

Nino Panagia

Laboratorio di Radioastronomia, Bologna, Italy
Laboratorio di Astrofisica Spaziale, Frascati (Roma), Italy

1. INTRODUCTION

In talking about infrared emission from HII regions, I am supposed to review the observational results and discuss possible theoretical interpretation. Recent reviews on the observational aspects are available in the literature (Wynn-Williams and Becklin, 1974; Olthof, 1975; see also the Proceedings of the 8th ESLAB Symposium on "HII Regions and the Galactic Centre", ed. by A.F.M. Moorwood, 1974, ESRO SP-105, and the Proceedings of the EPS Symposium on "HII Regions and Related Topics", ed. by T.L. Wilson and D. Downes, 1975, Lect. Notes Phys. 42, Springer-Verlag). Therefore, I shall limit myself to a brief summary of the most relevant observational information (Section 2) and instead will devote more space to the problem of the location and the nature (properties and amount) of the dust which is responsible for the infrared emission (Sections 3 and 4).

2. OBSERVATIONAL INFORMATION

2.1. The Spectral Distribution

The IR Spectrum is very similar for practically all HII regions, with only a few exceptions (Harper, 1975; Clegg et al. 1976). There is no doubt that it is a real continuum with an overall shape characteristic of thermal emission at low temperature (e.g. Wynn-Williams and Becklin, 1974; Ward et al. 1976). In the range 3 μm - 1 mm, the observed emission is greatly in excess relative to that expected for the gas. These arguments point to the dust as responsible for the IR radiation. The fact that the observed

G. G. Fazio (ed.), Infrared and Submillimeter Astronomy, 43-54.

absorption for wavelengths shorter than 2.2 μm may vary strongly from source to source (e.g. Wynn-Williams et al. 1972), although they are spatially close to each other, confirms the presence of large amounts of dust associated with individual HII regions. Therefore, in the following it will be given as established that the infrared emission is due to dust.

The spectrum is broader than that of an isothermal nebula, therefore a range of temperatures is required. For wavelengths longer than $\lambda_{peak} \simeq 50 - 100$ μm, the spectrum is steeper than a black body curve, which implies: i) the sources are optically thin at these wavelengths, and ii) the grain emissivity is a decreasing function of the wavelength, thus the grain size must be much smaller than 50 μm. In the range 2 - 20 μm, the spectrum can be fitted closely by a power law $F(\nu) \propto \nu^{-\gamma}$ with $\gamma \simeq 3.5 \pm 0.5$ (e.g. Frogel and Persson, 1974). This again implies a range of temperatures (Panagia, 1975). Also, in this spectral range the emitting nebulae are generally optically thin because the brightness temperatures are significantly lower than the black body color temperatures (Frogel and Persson, 1974).

In addition to a smooth continuum, there are some features either in emission (e.g. 9.7 μm band) or in absorption (9.7 μm and 3.1 μm bands) which give evidence for the presence of some particular dust components (silicate and ice; see Section 3). Also present are some atomic lines produced by fine-structure transitions of ionized elements, for example, [Ne II] λ 12.8 μm (e.g. Aitken and Jones, 1974) and [O III] λ 88.2 μm (Ward et al. 1975).

2.2 The Brightness Distribution

In the range 1 - 20 μm (hereafter the range 1 - 3 μm will be called near infrared, NIR, whereas the term middle infrared, MIR, will denote the range 3 - 30 μm) several HII regions have been mapped with high angular resolution (1" - 10"). Generally, the brightness distributions at all NIR and MIR wavelengths are similar to each other (e.g. Wynn-Williams et al. 1972; Frogel and Persson, 1974), with few exceptions (e.g. W51 - IRS2, Wynn-Williams et al. 1974). Therefore, the temperature range implied by the spectrum cannot simply be explained as due to dilution of radiation from the central star. The NIR and MIR maps coincide very well with the corresponding radio maps. Thus, some dust must be intimately mixed with the ionized gas.

In the far infrared (30 - 300 μm, FIR) the angular resolution is somewhat poorer than at shorter wavelengths, typically 1' - 4' (however, cf. Harvey et al. 1976). Within the limits of the available angular resolution, it is found that most of the FIR radiation originates in a region which closely agrees both in position and shape with the radio counterpart, although it may be somewhat more extended (e.g. Alvarez et al. 1974; Emerson et al. 1973; Furniss et al. 1975; Harper et al. 1976). Therefore, the FIR emitting dust is physically associated with HII regions, i.e.: i) It is either within the ionized region or immediately surrounding it; ii) It is energized by the same source(s) which ionizes the gas.

In the submillimeter range (300 μm - 1 mm, SMI) the observations are rather difficult and relatively few regions have been observed. However, there is mounting evidence that SMI radiation mainly originates from dust in molecular clouds (Soifer and Hudson, 1974; Hudson and Soifer, 1976; Righini et al. 1976; Werner et al. 1975; Westbrook et al. 1976; Clegg et al. 1976). SMI emitting grains and molecules would share the same energy source(s) (protostars?) and possibly have comparable temperatures (20-30K).

2.3 Correlations between Infrared and Radio Data

Harper and Low (1971) found that the FIR fluxes of several HII regions were related to the corresponding radio fluxes by a simple proportionality. They interpreted this as evidence for absorption of Lyman continuum (Ly-c) photons by the dust in competition with the gas. Subsequent analyses of more numerous data (Petrosian, 1974; Pottasch, 1974; Panagia, 1974) lead to the same conclusion. We report here some preliminary results of a very recent study by Felli and Panagia (1976). In Fig. 1 the total infrared luminosity, L_{IR}, is presented against the observed flux of Ly-c photons, N_L, for 46 HII regions. The data are taken from Emerson et al. 1973; Furniss et al. 1974, 1975 and references quoted therein. The following points are noteworthy: i) All sources lie well above the curves defining the loci of ZAMS stars and ZAMS clusters, as adapted from Panagia (1973) and Natta and Panagia (1976b); ii) The infrared excess, defined as IRE = $L_{IR}/[N_L\, h\nu(Ly\text{-}\alpha)]$, is essentially independent of the luminosity with an average value of about 14 and a dispersion of a factor 1.7. The high IR excess indicates that a substantial amount of dust is mixed with the ionized gas and absorbs a considerable fraction of the Ly-c photons, so that the N_L values deduced from radio observations are lower by a factor 2-3 than the Ly-c flux emitted by the exciting stars. The fact that the IR excess is practically the same at all luminosities may imply that the most powerful sources consist of several smaller components which are still unresolved at FIR wavelengths. A confirmation of this possibility comes from the fact that the biggest radio components found in well resolved HII regions have an excitation parameter of only U $\sim$ 85 pc cm^{-2} (Felli and Panagia, 1974; Wink et al. 1975). As a cross-check on the absorption of Ly-c radiation by dust, we can examine Fig. 2, in which the MIR luminosity is plotted against the expected Ly-α luminosity (Felli and Panagia, 1976). Since it is established that MIR radiation comes from within an HII region, we are certain to deal with emission by dust mixed with the ionized gas. Also, it must

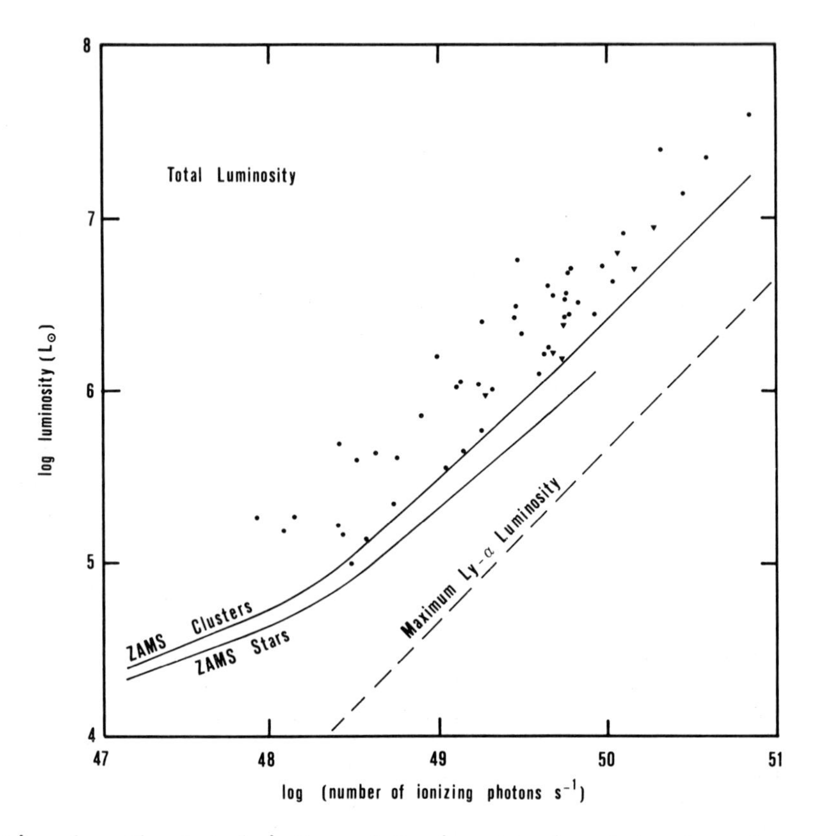

Fig. 1. The total infrared luminosity is plotted against the Ly-c photon flux for 46 HII regions. 3σ upper limits for some undetected regions are shown. For comparison, the curves of ZAMS clusters, ZAMS stars and the line $L_{IR} = L_{max}$ (Ly-α) are also shown.

be noted that the MIR luminosity is a lower limit to the emitted power by dust in HII regions, because it is possible that some FIR emitting dust is present within an HII region (Natta and Panagia, 1976a). We see that all sources emit significantly more than the Ly-α luminosity. Therefore, it is clear that the dust present within an HII region absorbs directly a considerable fraction of the stellar radiation and that absorption of Ly-α photons is not the dominant heating mechanism.

Another interesting correlation is that noticed by Churchwell et al. (1974) between the ratio of the abundances He^+/H^+ (derived from radio recombination line observations) and the infrared excess. They find that the ratio He^+/H^+ decreases as the IRE increases, and

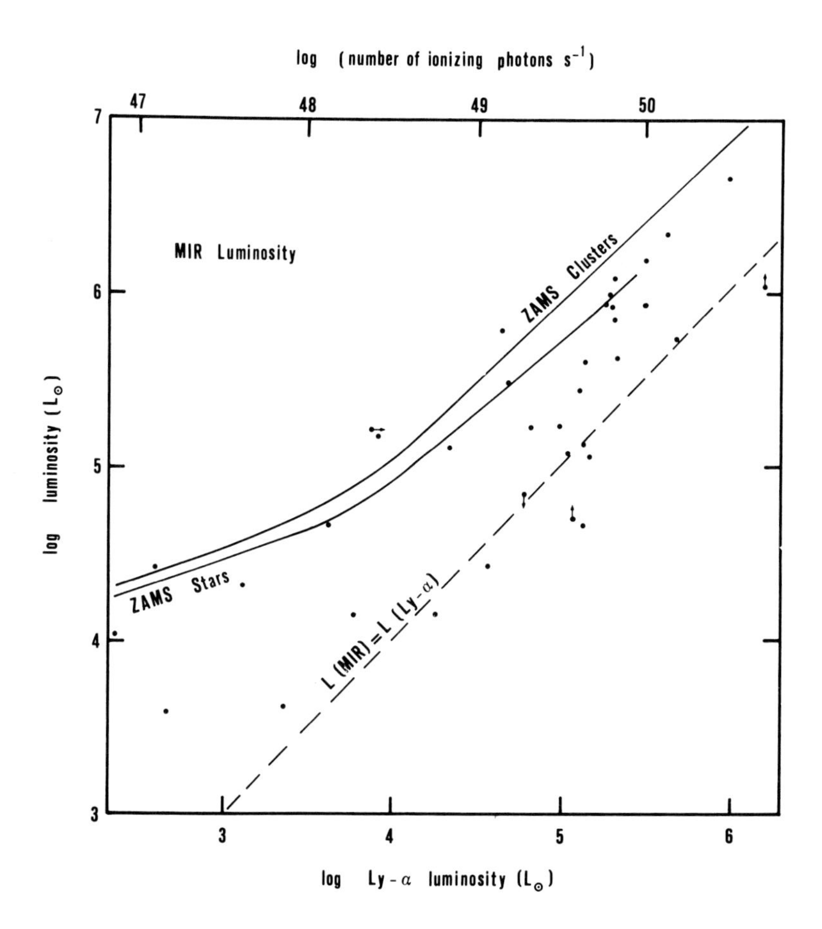

Fig. 2. The MIR luminosity is plotted against the Ly-α luminosity for 36 HII regions. The curves of ZAMS clusters, ZAMS stars and the line L = L(Ly-α) are also shown.

therefore He^+/H^+ decreases as the dust optical depth increases. The simplest explanation is that the effective opacity of dust in the He-ionizing continuum ($\lambda < 504$ Å, κ(He)) is higher than in the H-ionizing continuum ($504 < \lambda/\text{Å} < 912$, κ(H)) and that the differential absorption produces the apparent deficiency of He-ionizing radiation (Mezger et al. 1974). The required ratio of the two opacities would be $\kappa(\text{He})/\kappa(\text{H}) = 6 \pm 2$ (Smith, 1975; Mezger, this symposium). This value might have to be revised downward if the stellar continuum of O stars in the far UV is not so high as the model atmospheres would predict (cf. Panagia et al. 1976), but it should remain significantly greater than unity (say ~ 3). To produce such an increase of opacity in a small frequency interval (less than a factor 2) absorption by grains of small size ($\lesssim 100$ Å) and possibly with some sort of resonance at $\lambda \leq 500$ Å is required.

3. DISCUSSION

In this section we shall discuss possible interpretations of the observational data. Although quantitative estimates of various parameters will be given, one must bear in mind that the possible indeterminations are rather large and only the orders of magnitude are expected to be reliable.

The dust opacity in the Ly-c can be derived from a comparison $L_{IR} - N_L$ (e.g. Panagia, 1974). Adopting a spherical, constant density model to describe an HII region, the data presented in Fig. 1 yield an average opacity in the whole Ly-c (i.e. $\lambda < 912$ Å) of $\bar{\kappa} = 5.7 \times 10^{-22}\ n_H\ cm^{-1}$, n_H being the number density of hydrogen atoms. If it is assumed that $\kappa(He) = 4\ \kappa(H)$, then $\kappa(H) = 3.5 \times 10^{-22}\ n_H$ and $\kappa(He) = 1.5 \times 10^{-21}\ n_H$. These values represent lower limits to the actual opacities because any deviation from a uniform spherical model would raise them. For instance, adopting a spherical shell model with $\Delta R/R = \frac{1}{2}$ instead of a filled sphere would increase the quoted figures by a factor of about 2.5.

From the FIR fluxes it is possible to estimate a FIR optical depth and, by comparing it with the Ly-c optical depth, the absorption efficiency at $\sim 50\ \mu m$ can be inferred (Panagia, 1974; Natta and Panagia, 1976a). It has been found that $Q(50\ \mu m) \sim 0.01$. This implies an average size for the grains, which are responsible for the bulk of FIR emission as well as the bulk of the UV absorption (500 - 1500 Å), greater than 0.05 μm and a dust to gas ratio by mass $M_d/M_g > 4 \times 10^{-3}$. Since in order to account for selective absorption of He-ionizing photons grains with size ≤ 100 Å are required, this lower limit to the size of FIR emitting grains gives the first hint that several dust components must be present in HII regions.

Further information about dust properties in the FIR and SMI ranges is provided by some recent studies at 350 - 400 μm and 1 mm. It has been found that around 400 μm the dust absorptivity is approximately 20 times greater than what it could be for silicate and water ice grains but still about 10 times less than the maximum allowed by the Kramers-Krönig relations (Hudson and Soifer, 1976). In addition, it has been found that $Q(\lambda)$ decreases at least as fast as $\lambda^{-1} - \lambda^{-2}$ in the SMI range (e.g. Righini et al. 1976). Since these slopes for $Q(\lambda)$ have been determined by assuming that the spectrum is produced by an isothermal nebula, they should be regarded as lower limits to the actual values (Natta and Panagia, 1976a).

All the above results converge to form the picture of a dust

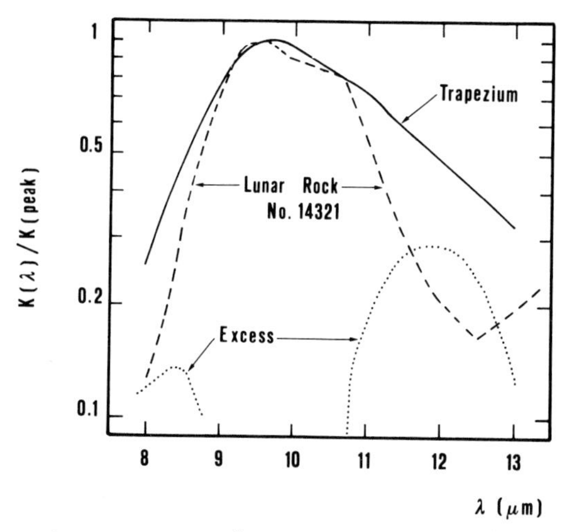

Fig. 3. The profile of the "10 μm" band observed in the Trapezium region (solid line) is compared with that computed for the lunar silicate No. 14321 (dashed line). The opacities are normalized to the peak value. Possible excesses around 8.5 μm and 11.8 μm are indicated (dotted curves).

component which dominates the UV absorption and the FIR emission, has a high FIR and SMI absorption efficiency, and has a relatively high abundance and large size. It cannot consist of either silicate or water ice and must be made of some other material (perhaps frozen molecules?).

On the other hand, evidence for the presence of water ice grains near, or even inside, HII regions is provided by the absorption band at 3.1 μm (e.g. Gillett and Forrest, 1973; Soifer et al. 1976). Silicate grains must also be present both within and around HII regions because several sources display a spectrum with an emission band centered at $\sim$ 9.7 μm attenuated by an absorption band with the same profile (Aitken and Jones, 1973; Gillett et al. 1975; Persson et al. 1976). Furthermore, recent measurements by Forrest and Soifer (1976) have revealed in the Orion nebula spectrum the presence of a band emission around 20 μm, which strengthens the silicate identification. However, even for these "silicate" bands, the observed profile is broader than that of any known silicate material. Looking at Fig. 3 we see that the intensity of the band observed in the Trapezium region (Forrest et al. 1975) is significantly in excess of that of the lunar rock No. 14321 (Bussoletti and Zambetta, 1976) especially around $\sim$8.5 μm and $\sim$ 11.8 μm. Such excesses may be explained by the presence of grains with a suitable band structure or, less well, by some extra source of continuous opacity. However, a mixture of different silicates does not seem to be able to reproduce the observed band profile, and some other constituent is needed. It is interesting to note that, in a region about 2' south-east of the Trapezium, spectra with local peaks at ∿8.7 μm and 11.2 μm have been observed (Becklin et al. 1976).

Also, in the NIR range, there is an indication that dust other than silicate contributes to the absorption. In fact, it has been found that $\tau(2.2\ \mu m) \simeq \tau(9.7\ \mu m)$ (Gillett et al. 1975; Soifer et al. 1976), whereas for any silicate the effective optical depth at 2.2 μm is from two to ten times lower than at 9.7 μm.

Therefore, we can conclude that several components are present in the dust associated with HII regions. The possibility of two coexisting dust components was first suggested by Lemke and Low (1972) to explain the extremely broad spectrum of M17. This suggestion has recently been repeated and reinforced by Harper et al. (1976) on the basis of more detailed data. Other direct and clear evidence for the presence of at least two components in one and the same region is provided by the observations of the bar in the Orion Nebula, approximately 2' south-east of the Trapezium (Becklin et al. 1976). It has been found that the spectrum requires a range of temperatures which cannot be due to dilution of radiation but which can be easily explained in terms of coexisting "hot" and "cold" dust.

From a theoretical point of view Natta and Panagia (1976a) have demonstrated that a model with two, or more, dust components can "naturally" account for all the observed features of HII regions as infrared sources. In particular, they have found that only with the presence of two components is it possible for dust to absorb a considerable fraction of the Ly-c flux and for the 10-20 μm color temperature to still remain almost constant across the HII region.

4. CENSUS OF THE DUST POPULATION

In the previous sections we have obtained some clear but fragmentary information on the dust associated with HII regions. Now we will present a unified picture which is derived by the assembly and organization of the available information with the aid of a little speculation when necessary. We propose a dust population which comprises four different components:

1) A first component (hereafter called D-dust) which dominates the UV absorption (500 - 1500 Å) and the FIR and SMI emission. It has a size of about 0.1 μm, a Ly-c absorption opacity of about $\bar{\kappa} \simeq 9 \times 10^{-22}\ n_H$, a high absorptivity in the IR ($Q \simeq 2\pi a/\lambda$ up to ∿100 μm) and an abundance relative to the gas of $M_D/M_{gas} \simeq$ $\simeq 5.5 \times 10^{-3}$. Its composition is unknown, but it may consist of frozen molecules, so that it should have a sublimation temperature of the order of 100K. Because of this, it can exist

only in the outer parts of, or outside, a compact HII region, thus its apparent UV opacity is two to three times lower than the actual one. Furthermore, the high abundance and relatively large size make it an important contributor to the visual and near infrared extinction. Scaling the absorption and scattering cross sections by λ^{-1} and λ^{-4}, respectively, for $\lambda > 3000$ Å, we estimate $\kappa_{ext}(5500 \text{ Å}) \simeq 5 \times 10^{-22}\ n_H$ and $\kappa_{ext}(2.2\ \mu m) \simeq 1 \times 10^{-22}\ n_H$. Then, the deficiency of dust found in the center of the Orion Nebula (Schiffer and Mathis, 1974) may be the effect of the evaporation of D-dust. In the most compact HII regions ($R < 5 \times 10^{17}$ cm) D-dust should be completely absent and only more resistent grains could survive. This may explain why K3-50 and DR21 have relatively small infrared excesses (< 6) but lower than normal He^+/H^+ ratios (0.07 and 0.05, respectively; Churchwell et al. 1974). Also, the fact that D-dust exists only at low temperature tends to tie the peak of the spectrum to some constant FIR wavelength, somewhat longer than the peak wavelength which corresponds to the sublimation temperature (Bollea et al. 1976). However, it is worth noting that even without invoking evaporation of D-dust the peak wavelength of the IR spectrum has been found to be approximately the same for very different conditions (cf. Natta and Panagia, 1976a).

2) A second component which is responsible for the increased opacity is the He-ionizing continuum. It may be identified with graphite, which seems to have a "resonance" around 25 eV (Taft and Philipp, 1965). The typical size is 0.01 μm and the opacity at 500 Å is about $\kappa(He) \simeq 1.5 \times 10^{-21}\ n_H$. This corresponds to a relative abundance by mass of 2×10^{-3}, which is about 40% of the cosmic abundance of carbon. Because this component is an efficient absorber only in the He-ionizing continuum, it can absorb about 10-15% of the total stellar energy, which is re-emitted mostly in the range 3-30 μm. It is interesting to note that these parameters for graphite are just those required for explaining the interstellar absorption bump peaking at 2175 Å (Savage, 1975).
3) Silicate dust is undoubtedly present (see Section 3). In the ionized region the abundance by mass is about 1.3×10^{-3}, which implies that $\sim$50% of the silicon is tied up in grains. This abundance is consistent with the data of Frogel and Persson (1974), Persson et al. (1976), and Gillett et al. (1975) when interpreted according to the prescriptions of Panagia (1975, 1976) and Natta and Panagia (1976a). The typical size is about 0.15 μm and the absorption opacity in the Ly-c $\sim 8 \times 10^{-23}\ n_H$. Thus, silicate dust absorbs, and re-emits in the IR (say 5-50 μm), about 20% of the total stellar energy as observed. At optical wavelengths the expected extinction is about $1.3 \times 10^{-22}\ n_H$ (nearly pure scattering).
4) Water ice is also present, mainly outside the HII region where the grain temperature is lower than 100K. Its abundance is about 1/4 that of silicate, as deduced from the optical depths in the 3.1 μm and 9.7 μm bands. If the same abundance is taken

for silicate in the cold parts of a nebula as is taken in the ionized region, the ice abundance turns out to be M(ice)/M(gas) $\sim 4 \times 10^{-4}$. Since ice should be present in the form of a coating of other grains, the size may be about 0.2 μm.

Other constituents may also be present (e.g. silicon carbide, carbonates, etc.), but they would not be important in determining any gross property of the IR emission or of the UV and IR absorption.

Of course, this proposed picture may not be the only one possible. However, it is interesting to note that not only is it able to explain "naturally" all major features of the infrared emission from HII regions, but, quite amazingly, it also provides a smooth fit to the properties of the dust in interstellar space. In fact, the total abundance by mass of the four listed components is M(dust)/M(gas) = 9.2×10^{-3}; the extinction expected in the visual is $\kappa_{ext}(V) \simeq 6.6 \times 10^{-22}\, n_H$ (with an albedo A ≃ 0.3), mainly due to D-dust and silicate and so on. These figures are only slightly larger than those appropriate for dust in interstellar space (cf. Macchetto and Panagia, 1973; Jenkins and Savage, 1974). On the other hand, D-dust could be characteristic of cold molecular clouds in the sense that D-dust grains may preferably grow in dense molecular environments. It is conceivable that dust in interstellar space be, on the average, less rich in D-dust than matter associated with HII regions, because only a part of the interstellar matter is in the form of dense clouds. Therefore, it is reasonable to conclude that dust associated with HII regions is essentially the same as dust in interstellar space, apart from some "obvious" modifications, such as growth of grains in the cold regions and evaporation of volatile grains in the innermost and hotter parts.

I wish to thank a large number of people who have sent me data in advance of publication, most particularly J.A. Frogel, D.A. Harper, M. Harwit, P.G. Mezger, M. Rowan-Robinson, M. Simon, B.T. Soifer and M.W. Werner. Also, I must thank D. Bollea, M. Felli and A. Natta for useful discussions. Many thanks are also due to L. Baldeschi for drawing the figures and to B. Mandel for typing the manuscript.

REFERENCES

Aitken, D.K. and Jones, B., 1973, Astrophys. J., 184, 127.
Aitken, D.K. and Jones, B., 1974, M.N.R.A.S., 167, 11p.
Alvarez, J.A., Furniss, I., Jennings, R.E., King, K.J. and Moorwood, A.F.M., 1974, ESRO SP-105, p. 69.
Becklin, E.E., Beckwith, S., Gatley, I., Matthews, K., Neugebauer, G., Sarazin, C. and Werner, M.W., 1976, preprint.

Bollea, D., Cavaliere, A., Natta, A. and Panagia, N., 1976, in preparation.
Bussoletti, E. and Zambetta, A.M., 1976, Astron. Astrophys. Suppl., in press.
Churchwell, E., Mezger, P.G. and Huchtmeier, W., 1974, Astron. Astrophys., 32, 283.
Clegg, P.E., Rowan-Robinson, M. and Ade, P.A.R., 1976, to appear in Astron. J., 81, 399.
Emerson, J.P., Jennings, R.E. and Moorwood, A.F.M., 1973, Astrophys. J., 184, 401.
Felli, M. and Panagia, N., 1974, Mem. S.A. It., 45, 335.
Felli, M. and Panagia, N., 1976, in preparation.
Forrest, W.J., Gillett, F.C. and Stein, W.A., 1975, Astrophys. J., 195, 423.
Forrest, W.J. and Soifer, B.T., 1976, preprint.
Frogel, J.A. and Persson, S.E., 1974, Astrophys. J., 192, 351.
Furniss, I., Jennings, R.E. and Moorwood, A.F.M., 1974, ESRO SP-105, p. 61.
Furniss, I., Jennings, R.E. and Moorwood, A.F.M., 1975, Astrophys. J., 202, 400.
Gillett, F.C. and Forrest, W.J., 1973, Astrophys. J., 179, 483.
Gillett, F.C., Forrest, W.J., Merrill, K.M., Capps, R.W. and Soifer, B.T., 1975, Astrophys. J., 200, 609.
Harper, D.A., 1975, Lect. Notes Phys. 42 (Springer-Verlag) p. 343.
Harper, D.A. and Low, F.J., 1971, Astrophys. J. (Letters), 165, L9.
Harper, D.A., Low, F.J., Rieke, G.H. and Thronson, H.A., Jr., 1976, Astrophys. J., 205, 130.
Harvey, P.M., Campbell, M.F. and Hoffmann, W.F., 1976, Astrophys. J. (Letters), 205, L69.
Hudson, H.S. and Soifer, B.T., 1976, preprint.
Jenkins, E.B. and Savage, B.D., 1974, Astrophys. J., 187, 243.
Lemke, D. and Low, F.J., 1972, Astrophys. J. (Letters), 177, L53.
Macchetto, F. and Panagia, N., 1973, Astron. Astrophys., 28, 313.
Mezger, P.G., Smith, L.F. and Churchwell, E., 1974, Astron. Astrophys., 32, 269.
Natta, A. and Panagia, N., 1976a, Astron. Astrophys., in press.
Natta, A. and Panagia, N., 1976b, in preparation.
Olthof, H., 1975, Ph.D. Thesis, University of Groningen.
Panagia, N., 1973, Astron. J., 78, 929.
Panagia, N., 1974, Astrophys. J., 192, 221.
Panagia, N., 1975, Astron. Astrophys., 42, 139.
Panagia, N., 1976, in preparation.
Panagia, N., Macchetto, F., Natta, A. and Preite-Martinez, A., 1976, Astron. Astrophys., in press.
Persson, S.E., Frogel, J.A. and Aaronson, M., 1976, preprint.
Petrosian, V., 1974, in "Interstellar Dust and Related Topics" IAU Symposium No. 52, ed. by J.M. Greenberg and H.C. van de Hulst, p. 445.
Pottasch, S.R., 1974, Astron. Astrophys., 30, 371.
Righini, G., Simon, M. and Joyce, R.R., 1976, Astrophys. J., in press.

Savage, B.D., 1975, Astrophys. J., 199, 92.
Schiffer, F.H., III, and Mathis, J.S., 1974, Astrophys. J., 194, 597.
Smith, L.F., 1975, Lect. Notes Phys. 42 (Springer-Verlag) p. 175.
Soifer, B.T. and Hudson, H.S., 1974, Astrophys. J. (Letters) 191, L83.
Soifer, B.T., Russell, R.W. and Merrill, K.M., 1976, preprint.
Taft, E.A. and Philipp, H.R., 1965, Phys. Rev., 138, A 197.
Ward, D.B., Dennison, B., Gull, G.E. and Harwit, M., 1975, Astrophys. J. (Letters), 202, L31.
Ward, D.B., Dennison, B., Gull, G.E. and Harwit, M., 1976, Astrophys. J. (Letters), 205, L75.
Werner, M.W., Elias, J.H., Gezari, D.Y., Hauser, M.G. and Westbrook, W.E., 1975, Astrophys. J. (Letters), 199, L185.
Westbrook, W.E., Werner, M.W., Elias, J.H., Gezari, D.Y., Hauser, M.G., Lo, K.Y. and Neugebauer, G., 1976, preprint.
Wink, J.E., Altenhoff, W.J. and Webster, W.J., Jr., 1975, Astron. Astrophys., 38, 109.
Wynn-Williams, G. and Becklin, E.E., 1974, Pub. A.S.P., 86, 5.
Wynn-Williams, G., Becklin, E.E. and Neugebauer, G., 1972, M.N.R.A.S. 160, 1.
Wynn-Williams, G., Becklin, E.E. and Neugebauer, G., 1974, Astrophys. J., 187, 473.

RADIO OBSERVATIONS OF HII REGIONS AND SOME RELATED THEORETICAL WORK

P. G. Mezger and J. E. Wink
Max-Planck-Institut für Radioastronomie,
5300 Bonn 1, Auf dem Hügel 69
Federal Republic of Germany

1. INTRODUCTION

In this paper we refer to the whole complex of radio and IR sources associated with an O-star as HII region. Regions where specific ions exist will be referred to as He^{+}-, H^{+}- and C^{+}-regions.

From both observations and model calculations the following qualitative picture of the evolution of an O-star emerges. A protostellar cloud of density > 10^{4} H-atoms cm^{-3} starts to collapse and at an early stage an "embryo star" forms at its center. While this embryo star accretes mass it moves up the main sequence (MS). First an "inner dust cocoon" forms which absorbs all stellar radiation and reemits it in the IR; typical dust temperatures in this cocoon are 1000 K. Gas is ionized within this cocoon. Radiation pressure from the IR sweeps the dust grains outside this cocoon and forms a second, expanding "outer dust cocoon". Since the neutral gas is coupled to the dust by friction some gas is swept with the dust and a region of low gas density develops between inner and outer dust cocoon. The shell of neutral gas and dust outside the outer cocoon may contain more than half of the protostellar cloud. At this stage the O-star is seen as a near IR point source inside a more extended far IR source of typical color temperature 100 K. About one free-fall time has elapsed since the collapse of the protostellar cloud started; for the protostellar clouds considered here this time is of order of some 10^{5} yr. A compact HII region forms once the inner cocoon breaks up. First its size is much smaller than that of the far IR source. Once the ionization front reaches the outer dust cocoon the sizes become similar. As the ionization front sweeps through the outer cocoon into the shell of neutral gas the HII region expands and evolves as an ionization bounded HII region with steadily increasing mass. Outside the H^{+}-region a C^{+}-region forms. Once the

G. G. Fazio (ed.), Infrared and Submillimeter Astronomy, 55-75.

ionization front reaches the edge of the protostellar cloud the HII region evolves as if it were density bounded with constant mass, while the ionization front moves rapidly into the region of low gas density. The turnover frequency ν_t which separates opaque and transparent part of the radio spectrum moves towards lower frequencies while the radio flux ($\nu>\nu_t$) decreases rapidly with time (see, e.g. Mathews, 1969). At this stage the HII region ceases to be a strong far IR source since the dust absorption depth for UV-photons becomes too small. The remnants of the compact HII region(s) are seen embedded in an extended low-density HII region. The lifetime of the HII region as a conspicuous thermal radio source lasts on the average about (6-8)10^5 yr. Later, the HII region is still observable in the optical range with its ionizing star(s) becoming gradually visible. For details we refer to Mezger and Wink (1975), Mezger and Smith (1975), Krügel (1975, Lect. Notes Phys. 42, p. 110) and Krügel and Mezger (1975) and references therein; and to sect. IV. 1) of this paper.

Radio continuum observations together with IR-observations yield the following information:

1) Stellar parameters: The total IR luminosity L_{IR} which is $\lesssim L_*$, the stellar luminosity; thus by means of stellar models one can estimate from L_{IR} parameters such as the number of Lyman continuum (Lyc) photons N_c or the number of carbon photons N_{carb} ($912 \lesssim \lambda/\text{Å} \lesssim 1101$) emitted per sec by the ionizing star. From the radio flux density of a transparent HII region one derives N_c', the number of Lyc-photons absorbed per sec by the gas.

2) From the shape of the opaque radio spectrum (Olnon, 1975; Panagia and Felli, 1975) and from the shape of the IR spectrum (Wright, 1973; Krügel and Mezger, 1975; Natta and Panagia, 1976) one may derive information about the distribution of ionized gas and dust.

3) The radio brightness temperature or flux density of an HII region can be compared with the surface brightness of the Hα-line (Ishida and Kawajirii, 1968) or with the combined free-free and free-bound flux density in the near IR (Wynn-Williams et al, 1972) and the extinction due to dust inside and in front of the HII region can be estimated. (However the extinction values derived in this way are usually in disagreement with observations of radio absorption lines, which yield or allow to infer total hydrogen surface densities N_H, which can be converted into visual extinction A_V by assuming a constant A_V/N_H-ratio throughout the Galaxy. This discrepancy may be due to clumping of the gas.)(See also sect.III.2.).

Radio continuum observations are widely used for the interpretation of IR-observations. We therefore limit the review in Sect. II. to recent high frequency single dish observations (II.1) and aperture synthesis (SRT) observations (II.2). IR-observers probably are less familiar with radio recombination line observations and their application to the interpretation of IR-observations. Therefore more space is devoted in Sect. III to recent developments in this field.

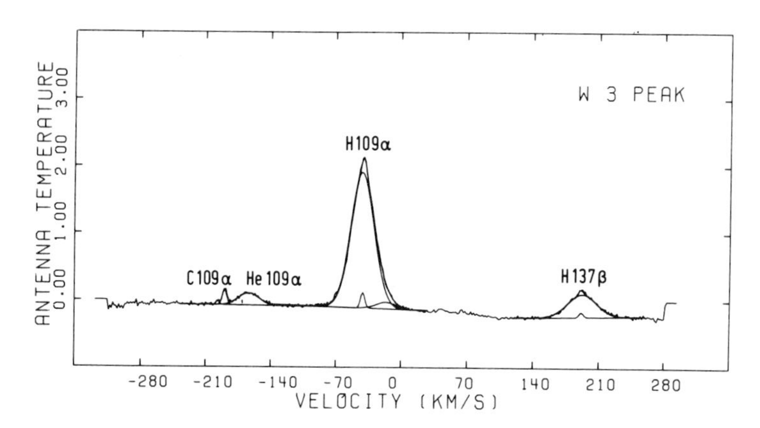

Figure 1: Recombination spectrum of W3, main component

Figure 1 shows the 109α recombination spectrum observed with the MPIfR 100-m telescope at the center of W3, main component. One recognizes the 109α recombination lines of H, He and C as well as the higher order transition H137β. From radio recombination lines one derives the following information:
4) From radial velocities, together with a model of galactic rotation, one estimates "kinematic" distances (III.1). The line widths and shapes yield information on large-scale motions, turbulence and local electron densities. (III.3).
5) The line-to-continuum ratio, especially of higher order transitions, allows estimates of the electron temperature.
6) He-lines in conjunction with H-lines yield information on the ionization structure and/or the abundances of these elements (see e.g. Churchwell et al, 1974 and III.2). (The IR Ne^{+} $\lambda 12.8\mu$ line yields similar information for Ne.)
7) The C-lines yield information on the physical state of the transition region between H^{+}- and H^{0}-region (see e.g. Walmsley,1975 Lect. Notes Phys. 42, p. 17) or on the C^{+}-region surrounding B-stars (III.4).

A special symposium on "HII regions and related topics" which summarizes radio observations of HII regions up to 1974 was held in Mittelberg, Austria, January 13 through 17, 1975. The proceedings of this symposium are published and will be referred to as (1975; Lect. Notes Phys. 42). In the following Sect. II (radio continuum) and III (radio recombination lines) we try to emphasize recent observational results not yet covered in these proceedings.

The organizers of this symposium also requested us to review recent theoretical work relevant to the interpretation of far IR observations. This is done in Sect. IV for the evolution of dusty HII regions (IV.1), the characteristics of dust (IV.2) and the effect of dust and variable element abundances on the observable parameters of HII regions (IV.3). Again those results are emphasized which are not yet covered in the symposium proceedings (1975; Lect. Notes Phys. 42).

II. RECENT RADIO CONTINUUM OBSERVATIONS

For the interpretation of far IR observations of special interest are radio continuum observations with high angular resolution and at high frequencies, which are reviewed in this section.

1) Single dish observations.
Standard catalogues of radio continuum (and H109α recombination line) parameters of galactic sources are: for the northern Milky Way the surveys by Altenhoff et al (1970) and by Reifenstein et al (1970). For the southern milky way the surveys by Goss and Shaver (1970) and by Wilson et al (1970). In addition there exists a catalogue of compact HII regions by Schraml and Mezger (NRAO 42.7-m, λ2cm, θ_A=2') which is often used for comparison with far IR-observations, and a survey of 168 optically identified HII regions by Felli and Churchwell (1972) made with the NRAO 91.5-m telescope (λ21cm, θ_A=10').

At present two rather complete surveys of the galactic plane are being prepared: For the northern part by Altenhoff, Downes, Schraml and Pauls (MPIfR 100-m, λ6cm, θ_A=2.6'); for the southern part by Caswell and Haynes (CSIRO 64-m, λ6cm, θ_A=4'). In addition various HII regions are being mapped in the wavelength range $6 \gtrsim \lambda/\mathrm{cm} \geq 1.35$ with the MPIfR 100-m telescope, the NRC 45.7-m telescope, the Haystack 36.6-m telescope, and the CSIRO 64-m telescope. Of special interest in the context of IR observations are radio observations in the wavelength range $13.5 \gtrsim \lambda/\mathrm{mm} \geq 1$, which links the far IR range with the more conventional (cm) radio range. Here the number of published observations compiled in Table 1 is still small.

Table 1

Recent mm-Observations of Galactic HII Regions.

λ/mm	Sources	References
3.3	DR21	Riegel and Epstein, 1968
3.5, 9.5	Sgr A, Sgr B2, W31, M17, W43, G45.5+01, G49.5-0.4, K3-50, DR21, NGC7538	Downes et al, 1970
3.5	Sgr B2, DR21, W3	Hobbs et al, 1971
9.5	W3, W49, W51, W75, Sgr B2, Orion B, NRAO584	Hobbs and Johnston, 1971
8.2	W3, Orion A, Orion B, W29, W31, W33, W38, W42, W48, W49, W51, K3-50, DR21, DR22, W43	Berulis and Sorochenko, 1973
2.14	DR21, W3, Orion A	Cogdell et al, 1975
13.5	RCW65, RCW87, RCW99, H2-3, H2-6	Brãz et al, 1975
8.2	Galactic center region	Berulis, 1976
1.23	W49, W51, Orion A, W3, W3OH, S159, K3-50, DR21, DR21OH, W75(N), SgrB2, Orion B, NGC2264, DR15, NGC7538,	Clegg et al, 1976
1	DR21, W75	Elias et al, 1976
1	DR21	Werner et al, 1975

Table 2

Compilation of Synthesis Observations of Galactic HII Regions

Source-Name	Wavelength [cm]	Resolution ["]	SRT	Lines observed	Reference
W3	75/21/6	80x91/23x26/7x8	CAM		Wynn-Williams 1971
	21	25x28	WBK	HI,H166α	Sullivan, Downes 1973
	6	40x66	OV	H_2CO	Fomalont, Weliachew 1973
	6	40x66	OV	H_2CO	Whiteoak et al. 1974
	11	6x7	NRAO		Webster, Altenhoff 1970a
	6	8x9	WBK	H109α	Wellington et al. 1976
	6	8x9	WBK		Harten in preparation
	11/3.7	6x7/2x3	NRAO		Wink et al. 1975
W3(OH)	50/21/6	56x63/24x27/∿5	WBK		Harten 1976
	21	25x28	WBK	HI,H166α	Sullivan, Downes 1973
	6	20x20	OV	H_2CO	Whiteoak et al. 1974
	6	7x8	CAM		Wynn-Williams 1971
	6	2x2.3	CAM		Baldwin et al. 1973
	11/3.7	6x7/2x3	NRAO		Wink et al. 1973
DR21	75/21	80x118/23x34	CAM		Ryle, Downes 1967
	11	6x9	NRAO		Webster, Altenhoff 1970a
	11/6	12x18/7x10	CAM		Wynn-Williams 1971
	6	7x11	WBK	H109α	Wellington et al. 1976
	6	2x3	CAM		Harris 1973
	11/3.7	8x9/2x3	NRAO		Balick 1972
W75(N)OH	21	23x148	CAM		Wynn-Williams 1971
	6	2x3	CAM		Harris 1974
W49	75/21	80x510/23x148	CAM		Wynn-Williams 1969a
	11	9x20	NRAO		Webster et al. 1971
	6	40x69	OV	H_2CO	Fomalont, Weliachew 1973
	6	6.5x42	CAM		Wynn-Williams 1971
	11/3.7	9x20/3x7	NRAO		Wink et al. 1975
W51-G49.5-0.4	11	8x16	NRAO		Turner et al. 1974
	11/6	12x50/7x27	CAM		Martin 1972
	6	40x65	OV	H_2CO	Fomalont, Weliachew 1973
G49.5-0.4	11/3.7	8x16/3x6	NRAO		Balick 1972
G49.5-0.4	11	<1	NRAO		Miley et al. 1970
G49.3-0.3	11/3.7	8x16/3x6	NRAO		Balick 1972
ORION A	11	7x35	NRAO		Webster, Altenhoff 1970b
	6	7x20	CAM/OV		Gull,Martin 1975,Lect.Not.Phys.42
	11/3.7	7x35/2.3x12	NRAO		Wink et al. in preparation
M17	11	9x40	NRAO		Webster et al. 1971
	11/3.7	9x40/3x14	NRAO		Wink 1974
M8	11	6x30	NRAO		Turner et al. 1974
	11/3.7	6x30/2x10	NRAO		Wink et al. 1975
NGC 7538	21	24x28	WBK		Israel et al. 1973
	11/6	11x13/2x3	CAM		Martin 1973
	11/3.7	6x7/2x3	NRAO		Wink et al. 1975
ORION B = NGC 2024	11	7x38	NRAO		Turner et al. 1974
	6	40x78	OV	H_2CO	Fomalont, Weliachew 1973
	11/3.7	7x38/3x13	NRAO		Wink et al. in preparation
W58-NGC 6857 -K3-50	75/21/6	80x145/23x42/2x3.6	CAM		Harris 1975
	50/21/6	58x105/25x45/7x13	WBK		Israel 1976
	11	12x22	CAM		Wynn-Williams 1969b
	11	6x10	NRAO		Wink et al. 1975
	11/3.7	6x10/2x4	NRAO		Lo 1974
W31	11	9x18/3x6	NRAO		Turner et al. 1974
	6	40x113	OV	H_2CO	Fomalont, Weliachew 1973
W43	11	9x38	NRAO		Turner et al. 1974
	6	40x78	OV	H_2CO	Fomalont, Weliachew 1973
ON-1	6	2x3	CAM		Harris 1974
	6	7x14	WBK		Winnberg et al. 1973
ON-2	6	2x3	CAM		Harris 1974
	6	7x12	WBK		Matthews et al. 1973
G75.84+0.4	11	7x9	NRAO		Turner et al. 1974
W33	11/3.7	8x30/3x10	NRAO		Balick 1972
NGC 2264south	11/3.7	9x20/3x7	NRAO		Balick 1972
NGC 2175	11/3.7	8x14/3x5	NRAO		Balick 1972
IC410	11/3.7	8x10/2x4	NRAO		Balick 1972
W28-A2	11	6x30	NRAO		Turner et al. 1974
G34.3-0.1	11	7x40	NRAO		Turner et al. 1974
G351.6-1.3	11/3.7	<1	NRAO		Broderick, Brown 1974
NRAO591	11	6x29	NRAO		Wink et al. 1975
G45.1+0.1	11	12x63	CAM		Wynn-Williams et al. 1971
G45.5+0.1	11/6	12x63/7x34	CAM		Wynn-Williams et al. 1971
NGC 7635	21	24x28	WBK		Israel et al. 1973
S157	21	24x28	WBK		Israel et al. 1973
IC1318 b+c	21	30x45	WBK		Baars, Wendker 1974
G79.3+1.3	21/6	30x45/6x10	WBK		Baars, Wendker 1974
G78.0+0.0	21/6	30x45/6x10	WBK		Baars, Wendker 1974
AFCRL-UAO#19	21/6	30x45/6x10	WBK		Wendker, Baars 1974
S254	21	25x80	WBK		Israel preprint
S255	21/6	25x80/7x23	WBK		Israel preprint
S256	21/6	25x97/7x23	WBK		Israel preprint
S257	21/6	25x97/7x23	WBK		Israel preprint
S258	21	25x80	WBK		Israel preprint
S266	6	7x27	WBK		Israel preprint
S269	21/6	25x102/7x30	WBK		Israel preprint
S271	21/6	25x115/7x34	WBK		Israel preprint
S272	21	25x115	WBK		Israel preprint
G192.58-0.04	21/6	25x97/7x23	WBK		Israel preprint
G192.60-0.04	21/6	25x97/7x23	WBK		Israel preprint
SGR A	21/18	22x37	*	HI, OH	Sandqvist 1973, 1974
	21/6	25x125/6.3x34	WBK/OV		Ekers et al. 1975
	18	195x195	OV	OH	Bieging 1976
	11/6	11x1000/6x540	CAM		Downes, Martin 1971
	6	40x154	OV	H_2CO	Fomalont, Weliachew 1973
	6	20x40	OV	H_2CO	Rogstad et al. 1974
	11/3.7	6x21/2x7	NRAO		Balick, Sanders 1974
	11/3.7	∿3/∿1	NRAO		Balick, Brown 1974
SGR B	21/18	22x37	*	HI, OH	Sandqvist 1973, 1974
	21	90x150	OV		Rougoor et al. 1974
	18	195x195	OV	OH	Bieging 1976
	6	40x148	OV	H_2CO	Fomalont, Weliachew 1973
	11/6	11x1000/6x540	CAM		Martin, Downes 1972

* = Lunar Occultations

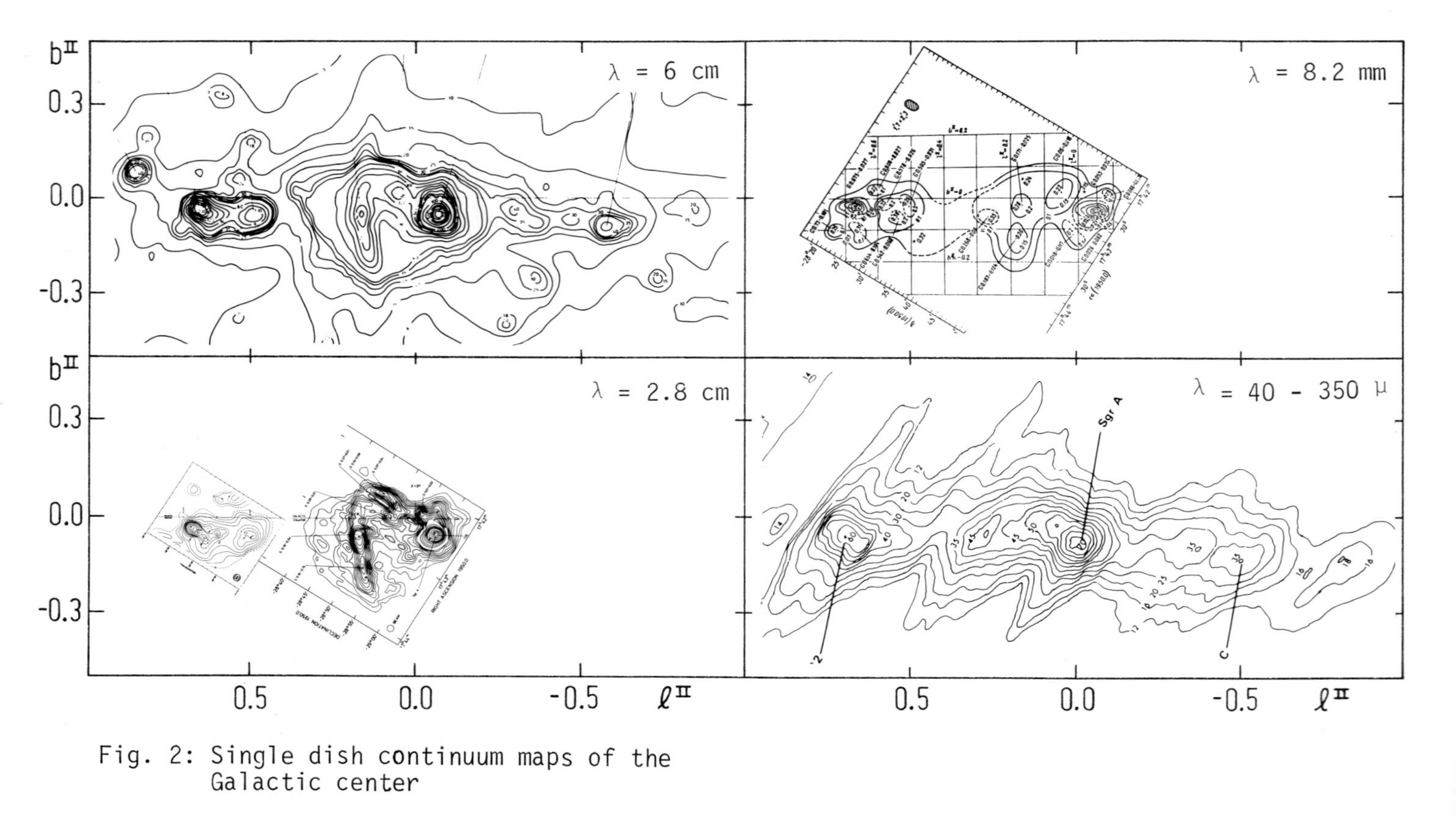

Fig. 2: Single dish continuum maps of the Galactic center

4

Friday
June
1982

8:00

8:30

9:00

9:30

10:00

10:30

11:00

11:30

12:00

1:00

1:30

2:00

2:30

3:00

3:30

4:00

4:30

5:00

5:30

June						1982
S	M	T	W	T	F	S
		1	2	3	4	5
6	7	8	9	10	11	12
13	14	15	16	17	18	19
20	21	22	23	24	25	26
27	28	29	30			

3

May						1982
S	M	T	W	T	F	S
						1
2	3	4	5	6	7	8
9	10	11	12	13	14	15
16	17	18	19	20	21	22
23	24	25	26	27	28	29
30	31					

July						1982
S	M	T	W	T	F	S
				1	2	3
4	5	6	7	8	9	10
11	12	13	14	15	16	17
18	19	20	21	22	23	24
25	26	27	28	29	30	31

As an illustration of this type of single dish continuum work we have compiled in Fig. 2 comparable maps of the galactic center region observed at λ6cm (Altenhoff et al, in prep.), λ2.8cm (Pauls et al, 1976; Downes, 1974), λ8mm (Berulis, 1976) and 350-40μ (Jennings, 1975, Lect. Notes Phys. 42, p. 137). At λ6cm still about 50% of the radiation maybe non-thermal while at λ8mm the main contribution is thermal free-free radiation from ionized gas. In the far IR one observes thermal radiation from dust grains. At present these maps are analysed to separate thermal and non-thermal radio components and to find out to which extent ionized gas and heated dust are spatially correlated (J. Schmidt, in prep.).

2) Observations with Synthesis Radio Telescopes (SRT's).
Most single dish observations have an angular resolution $\gtrsim$1'. Higher angular resolution is achieved by aperture synthesis techniques using synthesis radio telescopes (SRT's). These observations are of special interest for comparison with earth-bound high resolution IR-observations $\lambda < 40\mu$. As a service for IR-observers we have compiled in Table 2 all synthesis maps of galactic HII regions published to date.

The four SRT's which contributed to these observations are located in Westerbork, the Netherlands (= WBK), Cambridge, UK (= CAM), Green Bank, USA (= NRAO) and Owens Valley, USA (= OV). However, IR-observers not familiar with SRT's should be aware of the following problem. Representing the two-dimensional brightness distribution of a radio source by its spatial Fourier transform a SRT acts like a band-pass (as compared to a low-pass characteristic of a single dish telescope), whose low-frequency cut-off is determined by the closest configuration of the elements (= single telescopes) of the SRT. Large-scale structure of the radio source represented by Fourier components below the cut-off will be suppressed in the synthesis map. This is illustrated by the three maps of Orion A shown in Fig. 3.

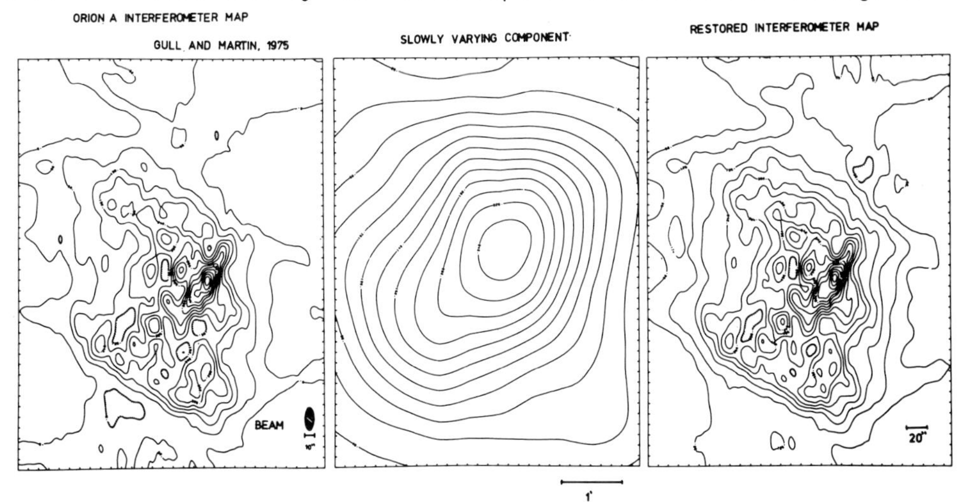

Figure 3: Radio maps of Orion A (from left to right): SRT map; single dish map; combined map.

The left-hand map has been obtained by Gull and Martin (1975, Lect. Notes Phys. 42, p. 369) by combining observations with the OV SRT and with the CAM SRT (λ6cm; 7.5"x20"). The center map, observed with the MPIfR 100-m telescope (Wink, unpublished, λ2.8cm, θ_A=76"), was corrected for the wavelength difference and the left-hand map, smoothed out to 76", was subtracted. The center map thus represents the extended background brightness distribution rejected by the SRT. The right-hand map is the superposition of left-hand and center map and represents the true brightness distribution of Orion A as would be seen at λ6cm with a single dish telescope of angular resolution 7.5"x20". (This map was prepared by D. Jaffe for comparison with optical Hα and Hβ observations). The main difference between left-hand and right-hand map are as expected the higher peak intensities in the latter map. In most cases both SRT and single dish observations of HII regions are available. In this case the ratio of the flux density contained in the SRT map to the flux density contained in the single dish map measures that fraction of the total flux density which is suppressed by the SRT. For example in W3, main component this ratio is close to unity, while in W3, southern extension this ratio is about 15% (Mezger and Wink, 1975). In general high resolution observations confirm the structure of HII regions first pointed out by Schraml and Mezger (1969) of compact components embedded in extended low-density regions. We refer to the symposium proceedings (1975, Lect. Notes Phys. 42) for detailed reviews of observations relating to the galactic HII regions Orion A and B (NGC2024), M17, DR21, W3, W49 and W51. In the same issue (p. 156) Habing discusses the structure of a number of HII regions based on SRT observations and its relation to the process of star formation. The distribution of the ionized gas in compact HII regions is by no means smooth. One observes density gradients, shell structures and local density variations such as shown for the Orion Nebula in Fig.3. In addition there is mounting evidence that the plasma in HII regions tends to clump. Defining a clumping factor Φ by n_e^2(clumps) = Φ n_e^2 (hom) with n_e (hom) the rms density determined in the usual way from the radio flux density and HPW of the radio source. The density in the clumps comes from strengths of optical forbidden lines, which are affected by collisional de-excitations (see also Sect. III.3). Observational values of Φ in the range 10 to 100 appear to be typical. Formation of substellar fragments in areas of star formation may cause these inhomogeneities.

Radio observations of HII regions in external galaxies are reviewed by Israel (1975, Lect. Notes Phys. 42, p. 288). Comparison of radio and optical observations show, that these HII regions contain large amounts of dust.

III. RECENT RADIO RECOMBINATION LINE OBSERVATIONS

In Sect. I, 4) through 7) are listed those characteristics of HII regions which are primarily determined by observations of radio recombination lines. Here we review recent results which appear of special relevance for the interpretation of far IR observations.

1) Kinematic Distances and the Galactic Structure

The most complete recombination line surveys of the galactic plane are the H109α surveys by Reifenstein et al (1970) and by Wilson et al (1970). With the MPIfR 100-m telescope all galactic sources with peak flux densities $S_5 \gtrsim 1$.f.u. have been searched for their H110α emission and H_2CO λ6cm absorption lines (Bieging, Downes, Wilson and Wink, in prep.). For the conversion of radial velocities into (kinematic) distances one usually uses the rotation curve derived by Schmidt (1965). For HII regions located closer to the galactic center than the sun there exists a distance ambiguity (referred to as "near" and "far" distance) which can be removed for example by observations of absorption lines against the free-free radio continuum of the HII region, which arise in the neutral gas located between HII region and the sun. One must realize that there are uncertainties in the rotation curve; the observational evidence of radial motions of the interstellar gas; and that the spiral structure using λ21cm kinematic distances does not agree very well with the spiral structure determined by direct distance measurements of, e.g. O-stars. Therefore kinematic distances should be considered as a first, crude approximation to the true distance of an HII region.

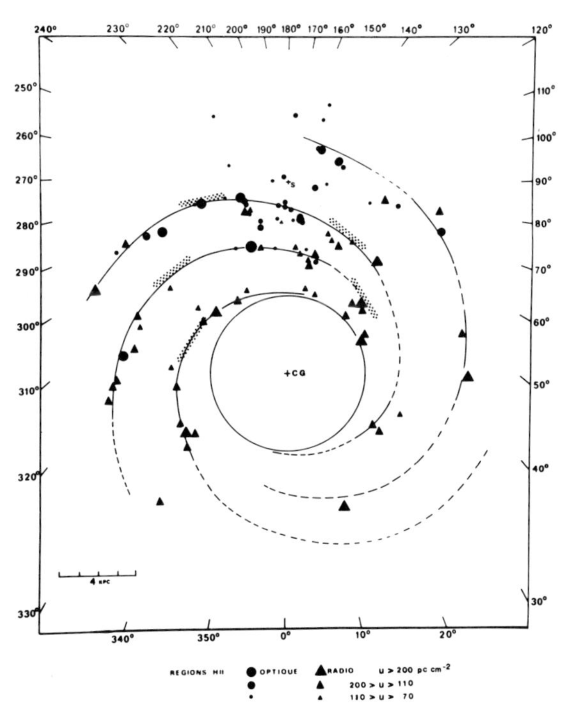

Figure 4: Spiral structure of the Galaxy as outlined by giant HII regions.

Fig. 4 shows the spiral structure of our Galaxy as outlined by giant HII regions, based on H109α surveys, optical observations and other radio spiral arm indicators. Although this four-arm structure derived by Yvonne Georgelin (1975a, b) may not yet be the final answer, it is certainly a much better approximation to the true spiral structure than earlier attempts based primarily on H λ21cm observations.

Apart from the main spiral arms (often referred to as "density wave spiral arms") giant HII regions are also found close to the galactic center. Many small HII regions are located in the interarm region, where they are often associated with cloud complexes referred to as "material arms". Thus the formation of O-stars (and hence of their associated IR-sources) is not limited to the main spiral arms where - according to estimates by Mezger and Smith (1975) - at present about 70% of all O-stars are formed; about 20% of the O-stars form in the interarm region and 10% within the inner 200 pc of the galactic center. These authors also estimate that 20% of all O-stars in the Galaxy are associated with conspicuous "radio HII regions" with rms electron densities of $\gtrsim$100 cm^{-3}. The percentage of strong IR sources associated with O-stars is even smaller, although in the very early stages IR-sources are not associated with observable HII regions. However, nearby B-stars, which are not hot enough to form HII regions may also be observable as far IR sources (see also sect.III.4).

2) The Ionization Structure of He

In HII regions with a He-abundance of y=0.1 (by number), which are ionized by stars O9 or earlier, He^+- and H^+-volumes should coincide, i.e. the quantity $R_0 = \int_{V(He^+)} n^2(H^+)dV / \int_{V(H^+)} n^2(H^+)dV$ should be unity. Churchwell et al (1974) surveyed 39 galactic HII regions and found ionized He-abundances y^+ ranging from 0.1 to less than 0.01 (close to the galactic center). Assuming y= 0.1 (which is probably correct for our Galaxy with the possible exception of the center region and outside 12 kpc) one derives R_0 from the relation $y^+=R_0 y$. Mezger et al (1974) plotted the quantity I_0-1 (with $I_0=L_{IR}/N_c' h\nu_\alpha$) against R_0. If dust absorbs Lyc-photons N_c' must decrease with increasing absorption depth τ_{Lyc}. With $\langle\sigma_{He}\rangle$ the absorption cross section for He-photons ($228 \leq \lambda/A \leq 504$), $\langle\sigma_H\rangle$ that for H-photons ($504 \leq \lambda/A \leq 912$) we define the quantity $a_0 = \langle\sigma_{He}\rangle/\langle\sigma_H\rangle$. If dust absorbs Lyc-photons selectively and if $a_0 > 1$ R_0 must decrease with increasing τ_{Lyc}. Such a correlation between I_0-1 and R_0 was found and a_0 was estimated. Jennings (1975, Lect. Notes Phys. 42, p.137) redetermined L_{IR} and N_c' using consistent observational material. In Fig. 5 is plotted I_0 versus R_0, using Jennings material for I_0 and the data by Churchwell et al (1974) plus some more recent, unpublished results of the 100-m telescope for R_0.

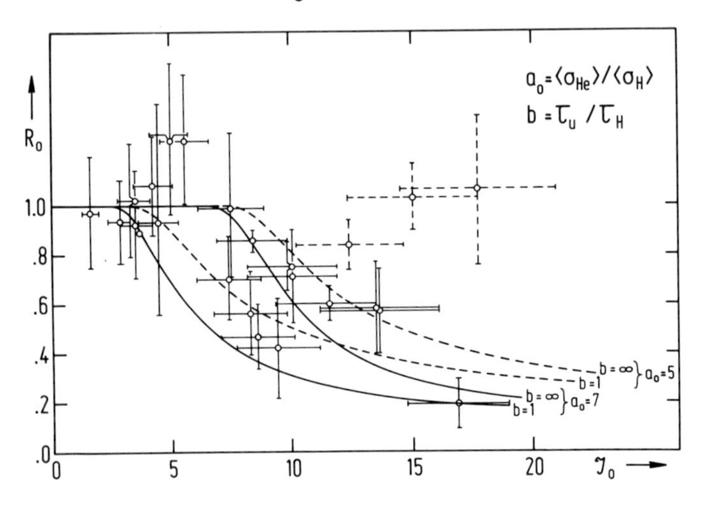

The curves represent new model fits by Mezger and Smith (in prep.). Considering all possible influences curves with parameters $a_0=6^{+1}_{-2.5}$ give the best fit.

Fig.5: Correlation between He^+-abundance and IR-luminosity, expressed in units of energy available in $L\alpha$-photons.

In model computations of the ionization structure of HII regions it is sometimes preferred to approximate selective absorption of dust grains by a power-law dependence of the absorption cross section on frequency, i.e. $\sigma_{Lyc}(\nu) = \sigma(\nu_1)(\nu/\nu_1)^{\alpha}$ with ν_1 the Rydberg frequency and $\alpha>0$. a_0 and α are connected by $a_0=0.37(4^{\alpha+1}-1.81^{\alpha+1})(1.81^{\alpha+1}-1)^{-1}$, yielding $a_0=2.1$ for $\alpha=1$ (the frequency dependence of small particles $2\pi a/\lambda<1$ with constant refractive index given by Mie theory). The value $a_0=6^{+1}_{-2.5}$ corresponds to an exponent $\alpha=2.5^{+0.2}_{-0.8}$. In addition to the parameter a_0 this model fit also yields absolute values of the absorption cross section of dust grains. For H-photons the absorption cross section of dust per H-atom is only about 40% of the corresponding extinction cross section in the visual (λ5500 A). For any plausible extrapolation of the extinction curve into the Lyc-range from observations made at $\lambda>912$ A in the general interstellar matter, this low absorption cross section implies very high values of the albedo $\Gamma>0.9$. However, this implication could be wrong for two reasons. Firstly, there is observational evidence (summarized by L. F. Smith, 1975, Lect. Notes Phys. 42, p. 175) that the physical characteristics of dust are different for HI and HII regions. Secondly, the absolute (but not the differential!) value of the absorption cross section of dust per H-atom as determined by Mezger et al (1974) depends on the clumping of the ionized gas. We designate the absorption depth derived on the assumption of a homogeneous plasma by τ_{Lyc} (hom). Three cases are possible: 1) The size of the HII region has been overestimated; 2) the plasma is distributed in random clumps; 3) the plasma forms a shell whose thickness is small as compared to its diameter. The relation between actual optical depth (i.e. for the "clumpy" plasma) and τ_{Lyc} (hom) is (Mathis, priv. comm.)

$$\tau_{Lyc}\text{ (clumpy plasma)} = \tau_{Lyc}\text{ (hom)}\cdot\begin{cases}\Phi^{1/6} & 1)\\ \Phi^{-1/2} & 2)\\ 3^{-1}\ \Phi^{-1/2} & 3)\end{cases}$$

While high resolution continuum observations should eliminate case 1), shell structures and a random clumping of the ionized gas may not be detected by direct observations. For a typical clumping factor of 30 (Orion A, Osterbrock and Flather, 1959), the estimated absorption cross section for H-photons would increase by a factor of five.

The most extreme example of low He^+-abundances is found in two giant HII regions close to the galactic center, G0.7-0.0 and G0.5-0.1, where earlier observations yielded values $y^+ <0.01$ to 0.02. Recently Brown and Lockman (1975) and Mezger and Smith (1976) reported the detection of He-recombination lines in these HII regions using large telescopes (high angular resolution) at high frequencies. Mezger and Smith suggest the low He^+ abundance is the result of selective absorption of Lyc-photons by dust and an increasing dust-to-gas ratio close to the galactic center, yielding very low values of R_0 ("geometric effect"). Brown and Lockman suggest a combination of collision broadening and preferential non-TE emission of H-lines in a partly ionized transition region between H^+- and H^0-region ("external maser effect"). Smith and Mezger (1976) critically reviewed all pertinent observational data. They feel (what a surprise!) that these data favor the geometrical effect. In addition and as first pointed out by Balick and Sneden (1976) (see also Sect. IV.3) an abundance increase in both Y and Z will also affect the ionization structure of HII regions through its effect on the opacity of the atmosphere of the ionizing star.

He radio recombination lines have been observed in one HII region outside the Galaxy, in 30 Doradus in the Large Magellanic Cloud. The more recent results by Huchtmeier and Churchwell (1974) yield values of 0.05 to 0.08 suggesting that in this giant HII region (which in our Galaxy is only comparable to the HII complex connected with the central region) He is neutral in part of the H^+-region.

3) Collision Broadening of Recombination Lines

The main reason why radio recombination lines were discovered relatively late was the fact that formulae developed originally for the collision broadening of optical lines were extrapolated into the radio regime of high n-values; this led to a gross overestimate of this effect. Only after the detection of H109α lines in very dense HII regions Griem (1967) derived the correct formulae for the broadening of radio recombination lines by electron collisions. Even on the basis of Griem's results one would expect easily observable collision broadening in high n transitions emitted from compact HII regions. This is not the case. For the Orion Nebula an explanation of this discrepancy between theoretical prediction and observation was first given by Brocklehurst and Seaton (1972). It hinges on the fact that Orion (as many other compact HII regions) has a high-density core surrounded by a lower-density envelope. At low frequencies (high values of n) collision broadening decreases the line intensity emitted by the dense central core, but at the same time non-TE effects ("internal maser effect") increase the intensity of the line emitted by the low-density envelope. The two effects act in such a way that for the observed "Gauss-fitted" line profile the line width (expressed in km s^{-1}) is approximately constant and that the line intensity does not deviate too much from its TE-value. This may be true for most core-halo HII regions, and may be the reason why only recently observations succeeded in at least a qualitative measurement of collision broadening in radio recombination lines, e.g. in DR21 (Pankonin, priv. comm.) and W3 (OH) (Hughes and Viner, 1976). Quantitative measurement of collision broadening may become an important tool for the determination of local electron densities in compact HII regions, which are inaccessible to optical observations. This is demonstrated by Churchwell, Terzian and Walmsley (1976) for the case of the planetary nebulae NGC7027.

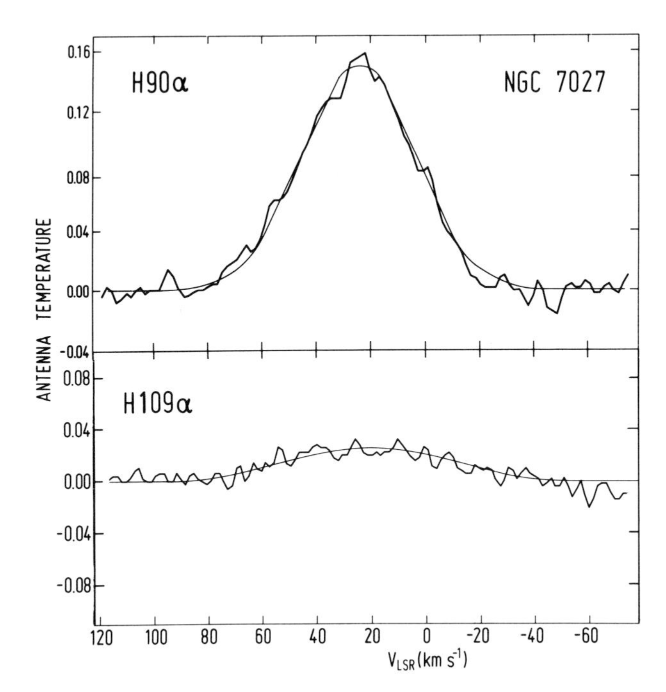

Figure 6: Recombination lines emitted by the planetary nebula NGC 7027

Figure 6 shows the H109α and the H90α lines emitted by NGC 7027. One clearly recognizes the effects of collision broadening in the H109α line. By model fitting (even here non-TE effects must be considered) with electron temperature and local density as free parameters most probable values of n_e=1.6 10^5 cm^{-3} and T_e=19000 K were derived. Compared with the rms electron density this corresponds to a clumping factor $\Phi \simeq 9$. High angular resolution observations at medium frequencies are required to extend this method to other compact HII regions. SRT recombination line observations such as reported by Wellington et al (1975, and ref. therein) of W3 and DR21 are promising.

4) Recombination Lines from C^+-regions

About one third of all radio HII regions observed with sufficiently high sensitivity show a carbon line in their recombination spectrum similar to the C109α line shown in Fig. 1. For earlier work we refer to the review by Walmsley (1975, Lect. Notes Phys. 42, p. 17). The C-lines associated with HII regions probably arise in the transition zone between H^+- and H^0-region with physical conditions $n_H \gtrsim 3\cdot10^4$cm^{-3} (in the following H-densities always refer to the total density of atomic and molecular H), $n_e \simeq 10$cm^{-3} and $T_e \simeq 100$ K. Due to absorption of UV-photons H_2 is dissociated in a large fraction, and CO and other molecules probably throughout the C^+-region. The Mg^+, S^+, Fe^+ and Si^+-regions are supposed to be even more extended than the C^+-region; these ions give rise to the "heavy element" recombination lines. At lower frequencies the contribution of stimulated emission becomes dominant and makes the C-lines more easily observable. Two facts make C-recombination lines ideal tracers of B-stars embedded in dense shells of dust and gas. The Lyc-photon flux drops by nearly three orders of magnitudes from stars of spectral type O4 to B0. Therefore B-stars have only very weak HII regions, if at all. The number of C-ionizing photons $N_{912-1101}$, on the other hand decreases in this stellar range only by one order of magnitude. Further, dust competes much more efficiently for absorption of C-ionizing photons than carbon. Walmsley's model computations show, that the C^+-emission measure which determines the intensity of the C-line emission of a resolved C^+-region is rather insensitive to the stellar UV-flux, but increases strongly with density. C^+-regions of B-stars embedded in dense neutral gas have first been detected for the Ophiuchus dark cloud by Brown et al (1974) and for NGC2023, a reflection nebulae by Knapp et al (1975). In the following we refer to the most recent observations of NGC2023 by Pankonin and Walmsley (1976)

Table 3

Stellar parameters and sizes of H^+- and C^+-regions of B-stars, computed for $n_H=10^4 cm^{-3}$ (Pankonin and Walmsley, 1976)

Spectral type	Log L* (erg s^{-1})	Log N_C (photons s^{-1})	R(H^+) (pc)	Log $N_{912-1101}$ (photons s^{-1})	R(C^+) (pc)
B0.5V	37.9	46.3	0.018	47.7	0.48
B1.5V	37.5	45.3	0.008	47.1	0.40
B4V	36.7	43.5	0.002	46.1	0.31
B6V	36.3	42.5	0.001	45.6	0.25

Both NGC2023 and the Ophiuchus dark cloud are well-known far IR-sources (e.g. Emerson et al, 1975; Fazio et al, 1976), and small compact HII regions have been discovered in the Ophiuchus cloud (Brown and Zuckerman, 1975). These observed characteristics can be used to derive the stellar parameters of the ionizing B-stars. In the case of NGC2023, the illuminating star of the reflection nebula is optically identified as B1.5V, in approximate agreement with the observed radio and IR-parameters.

An H^+-region is fully ionized and typical electron temperatures are 10^4 K. The dust in the HII region is heated by absorption of stellar radiation (including Lyc-photons) and radiation from the plasma such as trapped $L\alpha$-photons. The average temperatures of (ionized) gas and dust in the HII region differ typically by two orders of magnitude.

Figure 7, from Pankonin and Walmsley (1976) shows results of a model calculation for a C^+-region of density $n_H=10^4 cm^{-3}$ which is ionized by a B1.5 star. The heating of dust again is due to absorption of stellar radiation while photoionization of C and other elements is the prime source of heating for the gas, together with heating due to formation and dissociation of H_2. Photoelectric heating from dust grains (Watson, 1972) may be another important source of heating. The coupling between gas and dust temperature is small. Nevertheless the (observable) average temperature of dust and gas inside the C^+-region will be too different, while H_2 molecules can exist in the outer part of the C^+-region CO and other molecules

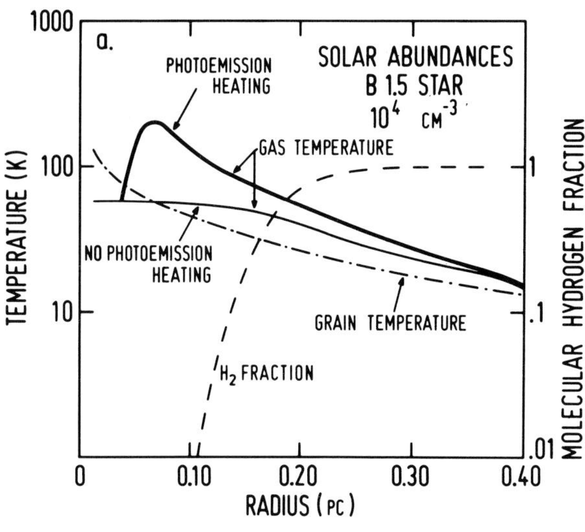

Figure 7: The physical state of a C^+-region

should be dissociated throughout the C^+-region. Thus C^+-regions should appear as "holes" in the observed brightness distribution of molecular radio lines. Observational evidence of such "holes" in the Ophiuchus dark cloud is claimed by Myers and Ho (1975). Of importance for future high resolution IR spectroscopy is the fact, that the prime cooling of the gas in the C^+-region is done by the following far IR lines: $C^+\lambda156\mu$, $O^0\lambda63\mu$, $Si^+\lambda35\mu$.

IV. RECENT THEORETICAL WORK

1) The Evolution of O-stars and Dusty HII regions

A qualitative picture of the evolution of O-stars and HII regions is given in sect. I. Some of the evolutionary stages outlined there have been treated quantitatively by Davidson and Harwit (1967), Larson (1969a, b), Larson and Starrfield (1971) and Kahn (1974). Recently Yorke and Krügel (1976) completed dynamical calculations for the envelopes of massive protostars starting from the point where protostellar clouds of 50 and 150 M⊙ respectively begin to collapse, until the point where the inner dust cocoon becomes tenuous enough so that a compact HII region can form. The evolution of a dust-filled compact HII region from an ionization bounded to a density bounded state was treated by Mathews (1969). It would be interesting to know how Mathews' results would be modified if he had used the initial conditions suggested by Yorke and Krügel's computations, i.e. a shell structure of neutral gas and dust with a dust front at its inner boundary.

In their computations Yorke and Krügel treated separate flows of gas and dust.(In the following quotation of numerical results we put the values referring to the 150M⊙ protostellar cloud in parenthesis). 3.2E5(1.5E5)yr after the collapse has started and shortly after a star-like nucleus has formed at the center of the protostellar cloud, an inner cocoon of radius 4.5E13(5E13)cm forms. Inside this cocoon even refractory graphite grains evaporate at an assumed melting temperature of 3650 K. Since the protostellar cloud is still opaque at near IR wavelengths, the inner cocoon is visible as an IR source of radius 5.8E14(1.6E15)cm and effective temperature 540(400)K. Radiation pressure from the inner dust cocoon acts on the dust component of the infalling material to which the neutral gas is coupled by

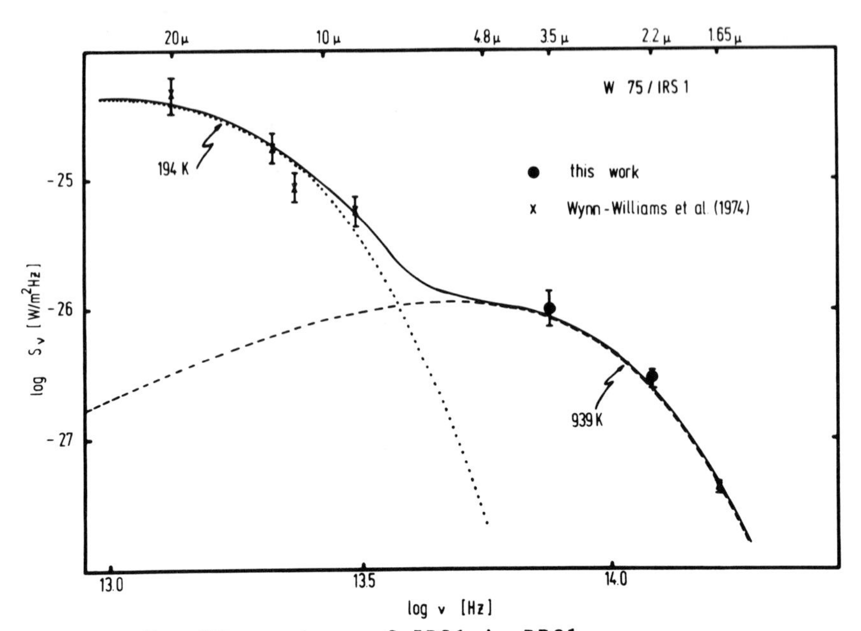

Figure 8: IR-spectrum of IRS1 in DR21

friction. Finally the mass flow is reversed and a second (outer) cocoon forms after 3.6E5(1.6E5)yr at a distance of 1E17(3E17)cm. The dust temperature of the outer cocoon is somewhat less than 100 K. In the subsequent evolution both the inner and outer dust cocoon expand, while the star still accretes mass from that part of the cloud which is inside the outer dust cocoon. Since the opacity of the inner cocoon steadily decreases its effective temperature will increase. After 3.7E5(1.9E5)yr the inner cocoon becomes optically thin to the near IR radiation and somewhat later to the stellar Lyc-radiation. At this point the quantitative computations stop. During the ∿1E3 yr in which the star accretes its final mass of 16(36)M☉, a very compact HII region forms inside the outer dust cocoon, which, at first, maybe ionization bounded. After the density has decreased sufficiently, the ionization front moves rapidly out into the second dust cocoon, until a (second) equilibrium Strömgren sphere has formed. Qualitatively, the following sequence of observable objects is predicted: First, the inner dust cocoon is seen as an IR source of color temperature ≃500 K whose spectrum drops very rapidly in the near IR due to extinction by cool dust. IRS 5 in W3 could represent this stage. Subsequently, the inner dust cocoon is seen as a hot IR point source (≃1000 K) embedded in an extended far IR source (≃100 K), which is due to radiation from the outer dust cocoon. IRS1 of DR21 (Fig. 8, Thum and Lemke, 1975) could represent this stage, although the absence of an observable compact HII region is probably due to the fact that the exciting star is of spectral type B. Once the inner cocoon becomes transparent to stellar Lyc-radiation, a very compact HII region forms inside the outer dust cocoon. Now one expects to observe a compact HII region of very small mass ($\lesssim$0.01M☉) in a much more extended far IR source. W3(OH) may represent this stage (Krügel and Mezger, 1975). Once the ionization front reaches the outer dust cocoon, one expects to see a shell-like compact HII region together with a far IR source with approximately equal size. W3(A) may represent this stage (Krügel and Mezger, 1975). The further evolution of the HII region and associated IR source is outlined in section I.

2) The Characteristics of Dust

In sect. III.2 we reported new results on the absorption characteristics of dust grains in the Lyc-range. These results corroborate with other evidence for selective absorption. While a strong increase of the absorption cross section of dust between 912 and 228 A helps to explain a number of observational results it remains an embarrassment for theorists constructing dust models. In addition, the problem of the high albedo values estimated for the far UV ($\gtrsim$1400A)(Witt and Lillie, 1973) and for Lyc (912 to 504 A) (Mezger et al, 1974) has been removed. Witt and Lillie (1976) have revised their earlier results and now estimate albedo values of $\gtrsim$0.6. Furthermore clumping of the ionized gas - as outlined in III.2 - will increase the absorption cross section of dust per H-atom estimated by Mezger et al (1974) and thus lower the dust albedo estimated for the Lyc-region.

Mathis et al (1976) fitted the observed extinction curve $0.11 \leq \lambda/\mu \leq 1$ by a least square fit with size distribution and chemical composition (within plausible restrictions) as free parameters. Graphite was found to be a necessary component, no ice mantles were needed for the fit and the derived particle size distribution can be approximated by $n(a) \propto a^{-3.3}$ within the size range $1 \lesssim a/\mu \lesssim 0.001$.

3) The Effect of Dust and Variable Element Abundances on the Ionization Structure of HII Regions

Recently a number of numerical computations of the ionization structure of HII regions (and other observable quanties) were made, including the effects of selective absorption of Lyc-photons by dust and variable element abundances. In these model computations selective absorption is either characterized by the parameter a_0 or the power exponent α. Both parameters can be easily related (see Sect. III.2). Unfortunately other initial conditions for the model calculations can not be so easily compared.

Balick (1975) computes the ionization structure of H, N, O, Ne and S for various kinds of frequency dependence of dust absorption and for variable dust-to-gas ratio. However, while the dust-to-gas mass ratio varies from $10^{-1.5}$ to 10^{-3} the element abundance stays constant. In a subsequent paper Balick and Sneden (1976) consider the effect of variable heavy element abundances Z in both HII region and ionizing star but now the effect of dust is neglected and the mass fraction of He is kept constant. In a more realistic model both the dust-to-gas ratio and that part of the He-abundance $\Delta Y(He)$ which results from element synthesis in stars should vary proportionally to Z. The most recent model calculations by Mathis (priv. comm.) allow the variation and combination of all these parameters. He predicts for example that in the case of selective absorption of

Lyc-photons by dust the electron temperature of HII regions should drastically decrease with increasing distance from the ionizing star. This could be checked e.g. by means of radio recombination lines. Two other important effects for the interpretation of IR- and radio observations are pointed out by Balick and Sneden: i) An increase in Z increases the cooling rate and thus decreases the electron temperature of an HII region (although quantitatively the effect appears to be overestimated). ii) An increase in Z changes the opacity of the atmospheres of O-stars in such a way that the stellar flux of photons above 35-40 eV is substantially reduced. This latter effect could in part be responsible for the low He^+-abundances of giant HII regions close to the galactic center (see Sect. III.2); and for the fact that a possible gradient of the He-abundance in galaxies has escaped detection (Mezger and Smith, 1976, Appendix A). Sarazin (1976) fits models to optical observations of HII regions in external Sc galaxies, in which both $\Delta Y(He) \simeq 3Z$, and the mass absorption coefficient of dust in the Lyc-region $k(\nu) = k(\nu_1)(\nu/\nu_1)^{\alpha}(Z/Z_{\odot})$ vary with Z, but apparently neglects the effect of Z on the stellar opacity. To explain the observations Sarazin finds that the dust parameters must be in the range $1 \lesssim \alpha \lesssim 3$ and $0.2 \lesssim k(\nu_1)/k_V \lesssim 2$, with k_V the visual extinction per H^+.

Of special interest for the interpretation of both the Ne^+ $\lambda 12.8\mu$ fine structure line and He-recombination lines are computations of the ionization structure of He and Ne. Mathis computations show that in dust-free HII regions there is only a very narrow range of effective stellar temperatures around T_{eff}=40 000 K were Ne^+ and He^+ coexist throughout the volume of the HII region. For higher values T_{eff} Ne would be mostly Ne^{++}, for lower T_{eff} He would be mostly He^0. However selective absorption of Lyc-photons by dust could alter this situation and for a given value T_{eff} increase the fraction of Ne^+ and decrease the fraction of He^+ in HII regions ionized by early O-stars. Selective absorption together with an overabundance of He and Ne relative to the vicinity of the sun may explain the observations of the He109α line (Pauls et al, 1974; Mezger and Smith, 1976) and the Ne^+ $\lambda 12.8\mu$ line (Aitkens et al, 1976; Wollmann et al, 1976) emitted from the compact HII region Sgr A West which is located at the center of our Galaxy. These observations are contradictory, since on the one hand the derived He^+- and Ne^+-abundances are consistent with solar system abundances. On the other hand Sgr A West, with diameter $\simeq$2pc, would have to be ionized by about hundred O9 stars to keep He and Ne singly ionized throughout the same volume.

We acknowledge critical remarks and helpful comments by D. Jaffe, H. Yorke, E. Krügel, J. Mathis, V. Pankonin and M. Walmsley, who also made available results prior to publication.

REFERENCES

Aitkens, D.K., Griffiths, J., Jones, B., Penman, J.M., 1976 MNRAS 174, 41P
Altenhoff, W.J., Downes, D., Goad, L., Maxwell, A., Rinehart, R., 1970 Astron. Astrophys. Suppl. 1, 319
Baars, J.W.M., Wendker, H.J., 1974 in F.J. Kerr and S.C. Simonson, III (eds) 'Galactic Radio Astronomy', IAU Symp. 60, p. 219
Baldwin, J.E., Harris, C.S., Ryle, M., 1973 Nature 241, 38
Balick, B., 1972 Ap. J. 176, 353
Balick, B., Sanders, R.H., 1974 Ap. J. 192, 325
Balick, B., Brown, R.L., 1974 Ap. J. 194, 265
Balick, B., 1975 Ap. J. 201, 705
Balick, B., Sneden, Ch., 1976 (preprint, Lick. Obs. Bull. No. 727)
Berulis, I.I., Sorochenko, R.L., 1973 Sov. Astron. 17, 179
Berulis, I.I., 1976 Sov. Astron. Lett.
Brãz, M.A., Jardim, J.O., Kaufmann, P., 1975 Astron. Astrophys. 43, 153
Brocklehurst, M., Seaton, M.J., 1972 MNRAS 157, 179
Broderick, J.J., Brown, R.L., 1974 Ap. J. 192, 343
Brown, R.L., Gammon, R.H., Knapp, G.R., Balick, B., 1974 Ap. J. 192, 607
Brown, R.L., Lockman, F.J., 1975 Ap. J. 200, L155
Brown, R.L., Zuckerman, B., 1975 Ap. J. 202, L125
Churchwell, E., Mezger, P.G., Huchtmeier, W., 1974 Astron. Astrophys. 32, 283
Churchwell, E., Terzian, Y., Walmsley, C.M., 1976 Astron. Astrophys. (in press)
Clegg, P.E., Rowan-Robinson, M., Ade, P.A.R., 1976 in press
Cogdell, J.R., Davis, J.H., Ulrich, B.T., Wills, B.J., 1975 Ap. J. 196, 363
Davidson, K., Harwit, M., 1967 Ap. J. 148, 443
Downes, D., Maxwell, A., Rinehart, R., 1970 Ap. J. 161, L123
Downes, D., Martin, A.H.M., 1971 Nature 233, 112
Downes, D., 1974 Proc. 8th ESLAB Symp. (A.F.M. Moorwood, ed.) p. 247
Ekers, R.D., Goss, W.M., Schwarz, U.J., Downes, D., Rogstad, D.H., 1975 Astron. Astrophys. 43, 159
Elias, J.H., Gezari, D.Y., Hauser, M.G., Werner, M.W., Westbrook, W.E., 1976 in press
Emerson, J.P., Furniss, I., Jennings, R.E., 1975 MNRAS 172, 411
Fazio, G.G., Wright, E.L., Zeilik II, M., 1976 Center of Astrophys. prepr. ser. No. 469
Felli, M., Churchwell, E., 1972 Astron. Astrophys. Suppl. 5. 369
Fomalont, E.B., Weliachew, L., 1973 Ap. J. 181, 781
Georgelin, Yvonne, 1975a Unpubl. thesis Université de Provence - Observatoire de Marseille, 1975b C.R. Acad. Sci, Paris, t. 280, series B p. 349
Goss, W.M., Shaver, P.A., 1970 Austr. J. Phys., Astrophys. Suppl. 14, 1

Griem, H.R., 1967 Ap. J. 148, 547
Harris, S., 1973 MNRAS 162, 5P
Harris, S., 1974 MNRAS 166, 29P
Harris, S., 1975 MNRAS 170, 139
Harten, R.H., 1976 Astron. Astrophys. 46, 109
Harten, R.H., 1976 in prep.
Hobbs, R.W., Johnston, K.J., 1971 Ap. J. 163, 299
Hobbs, R.W., Modali, S.B., Maran, S.P., 1971 Ap. J. 165, L87
Huchtmeier, W.K., Churchwell, E., 1974 Astron. Astrophys. 35, 417
Hughes, V.A., Viner, M.R., 1976 Ap. J. 204, 55
Ishida, K., Kawajiri, N., 1968 Publ. of the Astr. Soc. Japan 20, 95
Israel, F.P., Habing, H.J., de Jong, T., 1973 Astron. Astrophys. 27, 143
Israel, F.P., 1976 Astron. Astrophys. 48, 193
Kahn, F.D., 1974 Astron. Astrophys. 37, 149
Knapp, G.R., Brown, R.L., Kuiper, T.B.H., 1975 Ap. J. 196, 167
Krügel, E., Mezger, P.G., 1975 Astron. Astrophys. 42, 441
Larson, R.B., 1969a MNRAS 145, 271; 1969b MNRAS 145, 297
Larson, R.B., Starrfield, S., 1971 Astron. Astrophys. 13, 190
Lect. Notes in Phys. 42, Proc. of the Sympos. "HII regions and related topics" (T.L. Wilson and D. Downes, eds.), Springer Verlag
Lo, K.Y., 1974 Thesis Mass. Inst. Technology, Cambridge, Mass.
Mathews, W., 1969 Ap. J. 157, 583
Mathis, J.S., Nordsiek, K.H., Rumpl, W.M., 1976 (preprint)
Matthews, H.E., Goss, W.M., Winnberg, A., Habing, H.J., 1973 Astron. Astrophys. 29.
Martin, A.H.M., Downes, D., 1972 Astrophys. Letters 11, 219
Martin, A.H.M., 1972 MNRAS 157, 31
Martin, A.H.M., 1973 MNRAS 163, 141
Mezger, P.G., Smith, L.F., Churchwell, E., 1974 Astron. Astrophys 32, 269
Mezger, P.G., Wink, J., 1975 Mem. della Sociéta Astron. Italiana
Mezger, P.G., Smith, L.F., 1975 Proc. of 3rd European Meeting on Astronomy, Tblisi, July 1975 (in press)
Mezger, P.G., Smith, L.F., 1976 Astron. Astrophys. 47, 143
Miley, G.K., Turner, B.E., Balick, B., Heiles, C., 1970 Ap. J. 160, L119
Myers, P.C., Ho, P.T.P., 1975 Ap. J. 202, L25
Natta, A., Panagia, N., 1976 preprint
Olnon, F.M., 1975 Astron. Astrophys. 39, 217
Osterbrock, D.E., Flather, E., 1959, Ap. J. 75, 26
Panagia, N., Felli, M., 1975 Astron. Astrophys. 39, 1
Pankonin, V., Walmsley, C.M., 1976 Astron. Astrophys. (in press)
Pauls, T., Mezger, P.G., Churchwell, E., 1974 Astron. Astrophys. 34, 269
Pauls, T., Downes, D., Mezger, P.G., Churchwell, E., 1976 Astron. Astrophys. 46, 407
Reifenstein III, E.C., Wilson, T.L., Burke, B.F., Mezger, P.G., Altenhoff, W.J., 1970 Astron. Astrophys. 4, 357

Riegel, K.W., Epstein, E.E., 1968 Ap. J. 151, L33
Rogstad, D.H., Lockhart, I.A., Whiteoak, J.B., 1974 Astron. Astrophys. 36, 245
Rougoor, G.W., Högbom, J.A., Sandqvist, Aa., 1974 Astron. Astrophys. 33, 459
Ryle, M., Downes, D., 1967 Ap. J. 148, L17
Sandqvist, Aa., 1973 Astron. Astrophys. Suppl. 9, 391
Sandqvist, Aa., 1974 Astron. Astrophys. 33, 413
Sarazin, C.L., 1976 (preprint)
Schmidt, M., 1965 Galactic Structure (ed. A. Blaauw and M. Schmidt) Univ. of Chicago Press, p. 513
Schraml, J., Mezger, P.G., 1969 Ap. J. 156, 269
Smith, L.F., Mezger, P.G., 1976 Astron. Astrophys. (in press)
Sullivan, W.T., Downes, D., 1973 Astron. Astrophys. 29, 369
Thum, C., Lemke, D., 1975 Astron. Astrophys. 41, 467
Turner, B.E., Balick, B., Cudaback, D.D., Heiles, C., Boyle, R.J., 1974 Ap. J. 194, 279
Watson, W.P., 1972 Ap. J. 176, 103, 271
Webster, W.J., Altenhoff, W.J., 1970a Astron.J. 75, 896
Webster, W.J., Altenhoff, W.J., 1970b Astrophys. Letters 5, 233
Webster, W.J., Altenhoff, W.J., Wink, J.E., 1971 Astron. J. 76, 677
Wellington, K.J., Sullivan, W.T.III, Goss, W.M., Matthews, H.E., 1976 Astron. Astrophys. 47, 351
Wendker, H.J., Baars, J.W.M., 1974 Astron. Astrophys. 33, 157
Werner, M.W., Elias, J.H., Gezari, D.Y., Hauser, M.G., Westbrook, W.E., 1975 Ap. J. 199, L185
Whiteoak, J.B., Rogstad, D.H., Lockhart, I.A., 1974 MNRAS 169, 59 P
Wilson, T.L., Mezger, P.G., Gardner, F.F., Milne, P.K., 1970 Astron. Astrophys. 6, 364
Wink, J.E. Altenhoff, W.J., Webster, W.J., 1973 Astron. Astrophys. 22, 251
Wink, J.E., 1974 Thesis University of Münster
Wink, J.E., Altenhoff, W.J., Webster, W.J., 1975 Astron. Astrophys. 38, 109
Wink, J.E., Altenhoff, W.J., Webster, W.J., 1976 in preparation
Winnberg, A., Habing, H.J., Goss, W.M., 1973 Nature Physical Science 243, 78
Witt, A.N., Lillie, C.F., 1973 Astron. Astrophys. 25, 397
Witt, A.N., Lillie, C.F., 1976 UV-photometry from the OAO.XXV (preprint)
Wollman, E.R., Geballe, T.R., Lacy, J.H., Townes, C.H., Rank, P.M., 1976 Ap. J. 205, L5
Wright, E.L., 1973 Ap. J. 185, 569
Wynn-Williams, C.G., 1969a MNRAS 142, 453
Wynn-Williams, C.G., 1969b Astrophys. Letters 3, 195
Wynn-Williams, C.G., 1971 MNRAS 151, 397
Wynn-Williams, C.G., Downes, D., Wilson, T.L., 1971 Astrophys. Letters 9, 113
Wynn-Williams, C.G., Becklin, E.E., Neugebauer, G., 1972 MNRAS 160, 1
Yorke, H.W., Krügel, E., 1976 Subm. to Astron. Astrophys.

CONSIDERATIONS FOR THE INTERPRETATION OF INFRARED EMISSION FROM MOLECULAR CLOUDS

N. Z. Scoville

Department of Physics and Astronomy
University of Massachusetts
Amherst, Mass.

J. Kwan

Department of Earth and Space Sciences
State University of New York
Stony Brook, New York

The last years have seen a remarkable growth in observational data relating to the dense interstellar clouds composed of molecular hydrogen in which star formation occurs. Though molecular hydrogen itself has no observable radio frequency transitions, the gaseous component in these regions has been mapped using the millimeter wavelength rotational lines of trace molecules like CO. And where the associated dust grains have been heated to temperatures above 30°K, adjacent to either an HII region or an embedded "star", their emission has been detected in the far infrared. In the absence of such a heat source this emission would be unobservable at wavelengths shorter than 200μ. The far infrared measurements are, therefore, most useful in delineating regions of active star formation. The CO observations (reviewed in §5) indicate that the gas is also heated in the vicinity of the luminous sources. The dust grains and their infrared emission are, therefore, also critical to the energetics and energy transfer in molecular clouds. In this review I would like to describe an analysis we have recently made of the infrared radiative transfer in these molecular cloud sources (Scoville and Kwan 1976). We hope that this study not only aids in interpretion of the infrared data but also may shed light on the relationship between the dust and gas in these fascinating regions.

As a model we considered an opaque, spherically symmetric

G. G. Fazio (ed.), Infrared and Submillimeter Astronomy, 77-88.

cloud of dust grains surrounding a source of luminosity L_*. A main objective of our research was a thorough search for characteristics in the calculated model spectra which could delimit the unknown source parameters. (For example, we found that the spectrum peak must always occur at a wavelength λ for which the dust optical depth $\tau_\lambda \leq 3$). For this reason we restricted the model parameters to very simple forms though a second phase of the work would be to model the observations of the Kleinmann-Low nebula. In this spirit we investigated models where the dust density n_d fell off as r^0, r^{-1}, or r^{-2}. The grain emission (and absorption) efficiency was taken as a three part power law in λ. $\varepsilon_\lambda \propto \lambda^0$ for $\lambda < 1000\text{Å}$; $\varepsilon_\lambda \propto \lambda^{-1}$ for $1000\text{Å} < \lambda < 20\mu$; and $\varepsilon_\lambda \propto \lambda^{-2}$ for $\lambda > 20\mu$. This crude form is consistent with expectations for H_2O and silicate grains (Irvine and Pollack 1968 and Knacke and Thomson 1973). Once the dependence of dust opacity on radius from the star and on wavelength have been adopted, the only remaining critical parameter is the absolute value of the optical depth at 100μ, $\tau_{100\mu}$ measured from the outer radius of the dust shell through to the star. The adopted inner and outer radii at which one truncates the dust shell are easily seen to be unimportant provided the cloud is not optically thin ($\tau_{100\mu} \geq 0.1$) and provided the outer layers do not contribute substantial optical depth respectively.

1) Dust temperatures

At each radius in the dust shell the grain temperature is calculated from the condition of radiative equilibrium in the radiation of both the star and the other dust. Neglecting for the moment the secondary radiation, the equilibrium implies that

$$\frac{L_*}{4\pi r^2} \pi a^2 \varepsilon_{\lambda_{T_*}} = 4\pi a^2 \sigma T_d^4 \varepsilon_{\lambda_{T_d}} \tag{1}$$

and, therefore,

$$T_d \propto \left(\varepsilon_{\lambda_{T_*}} \frac{L_*}{r^2} \right)^{\frac{1}{5} \rightarrow \frac{1}{6}} \tag{2}$$

for $\varepsilon_{\lambda_{TD}} \propto \lambda^{-1 \rightarrow -2}$. In this simple case it is a well known result that the dust temperature will vary as the source luminosity $L_*^{0.2}$ and will fall off with radius as $r^{-0.4}$.

In the clouds of high optical depth which we model, there are two important modifications which we find largely counterbalance each other leaving at least the qualitative form of Equation (2) unchanged. Though the primary stellar radiation is severely attenuated by intervening dust between the star and the radius r, this deficit is largely made up by the subsequent reradiation. Since the absorption efficiency increases with decreasing wavelength, the most important contribution to the secondary heating at a given radius is almost always from grains more interior to r which have higher temperatures. Thus it is possible to visualize the changes occurring in Equation (2) when we move to optically thick clouds as a corresponding decrease in the effective $\varepsilon_{\lambda T_*}$ caused by degradation to longer wavelength of the original luminosity.

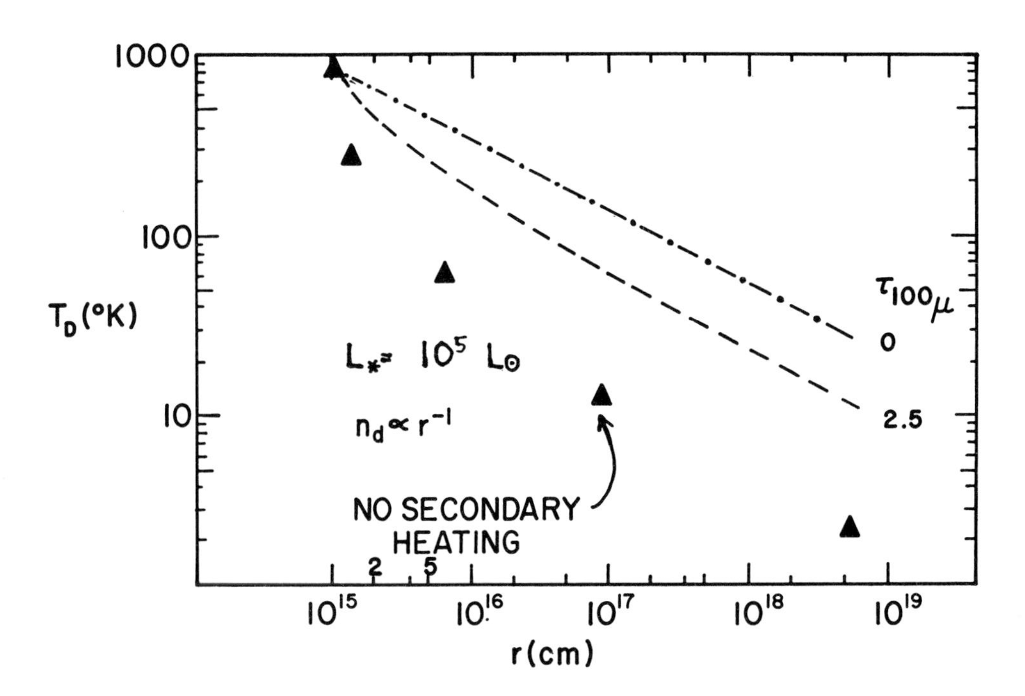

Figure 1: The decrease in dust temperature away from the central star for models with density varying as r^{-1} for optical depth $\tau_{100\mu} = 0, 2.5$. The importance of heating by dust reradiation may be judged from the triangles which are the temperatures calculated with only the heating from the attenuated starlight included, using the model with $n_d \propto r^{-1}$ and $\tau_{100\mu} = 2.5$.

The first figure illustrates the run of dust temperatures calculated in different degrees of sophistication for a single set of model parameters ($L_* = 10^5 L_\odot$ in order to match the total observed luminosity of the Kleinmann-Low nebula, $n_d \alpha r^{-1}$, and $\tau_{100\mu} = 2.5$). The upper curve results when both attenuation and reradiation are neglected in considering the radiative equilibrium; this is the behavior predicted in Equation (2). When just the attenuation between the star and radius r is also considered, the temperatures (Δ) are drastically decreased. But with inclusion of the reradiation term, the original run of temperatures is largely recovered even in this very high opacity model.

Since the grain temperature has only a 1/5'th root dependence on the color temperature of its incident radiation, the observed spectrum from an optically thick cloud will have lost all memory of the original stellar temperature T_*.

2) The spectrum

Once the dust temperature distribution had been determined, the emitted spectrum S_ν was calculated from the model as would be observed in circular diaphragms of radius ψ_{ap} centered on the source at the distance of the Orion nebula. This convolution calculation is unfortunately necessary before a meaningful comparison can be made with observations.

Figure 2 shows these spectra for two models differing only in that one is quite optically thin while the other is very thick. Two notable generalities which are true of all the models may be seen in these spectra. First: a characteristic property of optically thin spectra is that the spectrum peak will always shift to higher frequency when a smaller diaphragm is employed. And this shifting will continue only until the optical depth at the peak becomes ≈ 3. Thus, the optical depth at the peak in any of these models will always be ≤ 3. Secondly, we have found that optically thin spectra will have a power law form on the short side of the peak, and thick spectra will exhibit an exponential decay there. The index of the power law ν^{-3}, $\nu^{0.8}$, or $\nu^{1.4}$ in the thin spectra will moreover provide a clue to the density distribution r^{0}, r^{-1}, or r^{-2} respectively. (The first of these seems similar to spectra of several near infrared sources found in HII regions.)

Measurements at wavelengths longer than the peak of emission yield the form of the dust emissivity law there, but more importantly they also provide a unique handle to the dust density distribution. For grain temperatures occurring even at large radius in these sources, the Rayleigh-Jeans approximation is satisfactory at $\lambda > 300\mu$. Then since the calculated dust temperature varies slowly with radius $\alpha r^{-0.4}$), any major variation of the long wavelength intensity as one maps out from the center of a given source must be mostly due to optical depth or column density changes rather than temperature.

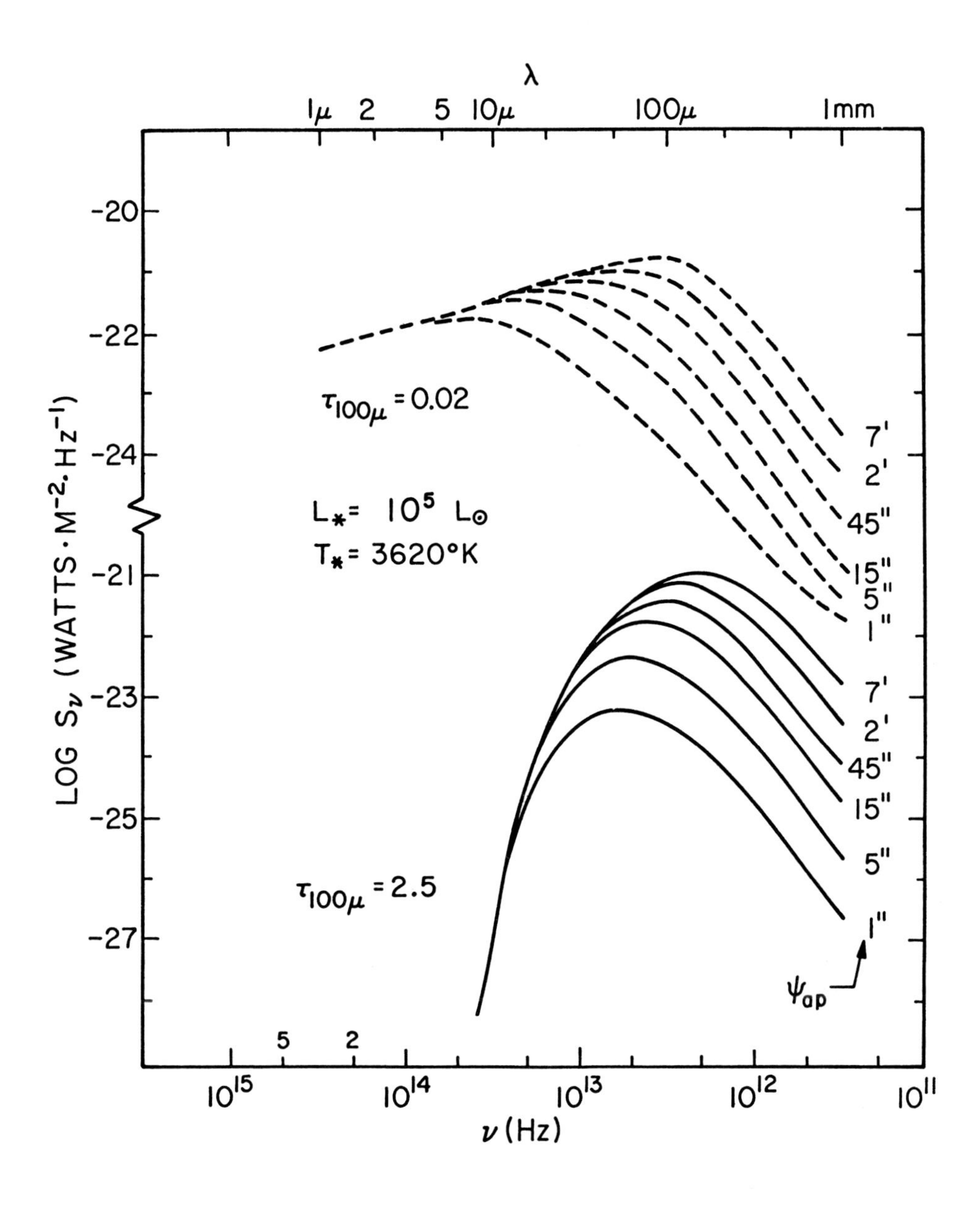

Figure 2: Spectra of optically thin and thick models with $n_d \propto r^{-1}$ and observed with various apertures. The aperture sizes are for a distance of 500 pc. Note that the optically thin source has a power law spectrum in the near IR and that the wavelength of peak emission is redshifted for larger apertures.

3) A model for the Kleinmann-Low nebula

Using the considerations above to match the calculated models to observations of the Kleinmann-Low nebula, the range of possible

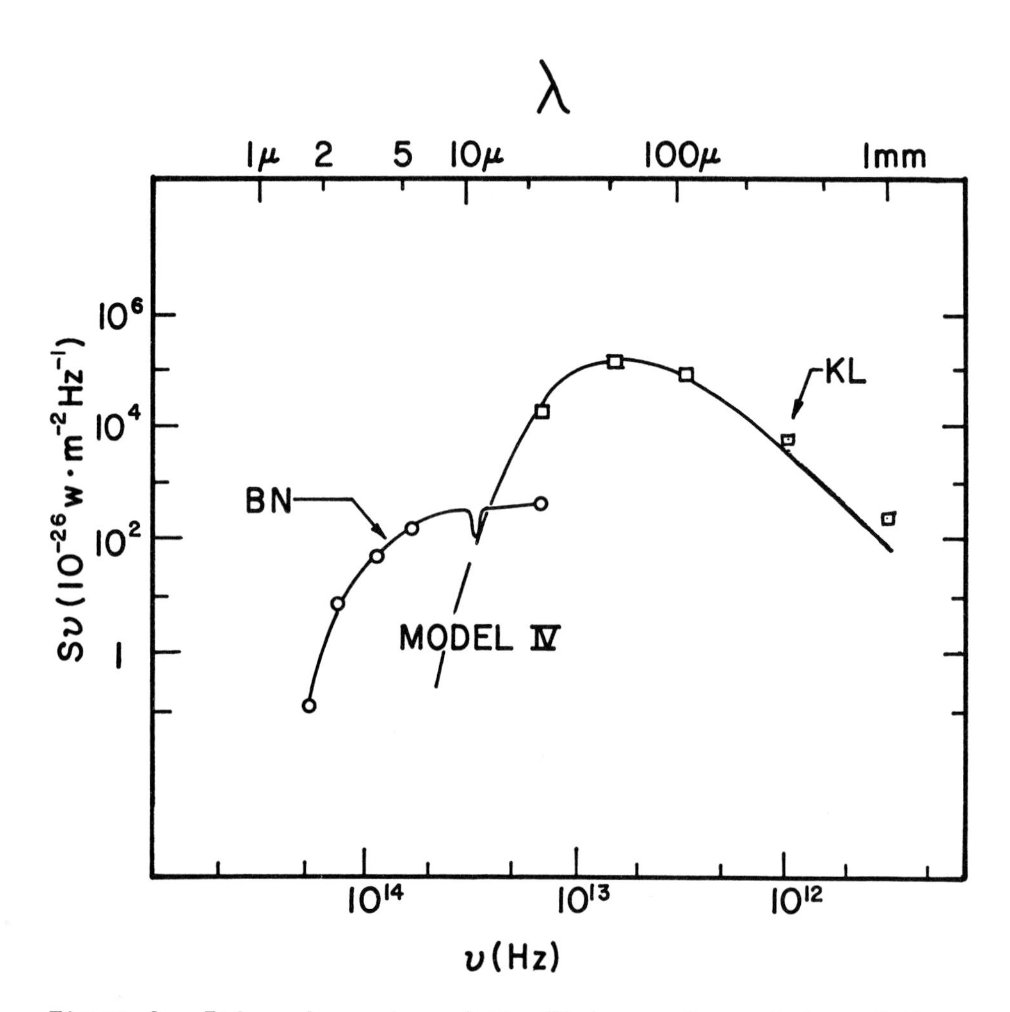

Figure 3: Infrared spectra of the Kleinmann-Low nebula and the Becklin-Neugebauer object as measured in a 1' beam at $\lambda \geq 20\mu$. At $\lambda \leq 20\mu$ the flux densities are obtained by integrating high resolution maps over a 1' beam, and thus we hope to have omitted low surface brightness emission excited by the HII regions. Data are from Becklin et al. (1973), Gillett and Forrest (1973), Werner et al. (1975), Harper et al. (1972), Gezari et al. (1974), Werner et al. (1974), Rieke, Low, and Kleinmann (1973). The spectrum (inside a 1' aperture) from the model with $\tau_{100\mu} = 0.5$, $n_d \propto r^{-1}$, and $L_* = 10^5 L_\odot$ has been plotted over the KL data.

source parameters may be restricted. First, we use the $\lambda = 1$ mm map of Harvey et al. (1974) to infer a density gradient of $r^{-1.5}$ at $r > 2 \cdot 10^{17}$ cm. (Based upon molecular line observations, an even steeper density gradient is likely at $r < 10^{17}$ cm).

Secondly, from this long wavelength tail of the observed spectrum $S_\nu \propto \nu^{3 \to 3.5}$, we infer an emissivity law of $\nu^{1.75}$. That the emissivity law index is not precisely the spectrum index-2 is due to minor departures from the Rayleigh-Jeans approximation (see Scoville and Kwan 1976).

Finally, we estimate that the source becomes optically thick at $\lambda \approx 70\mu$. This model is shown in Figure 3. The discrepancy between the model and the observed $\lambda = 1$ mm flux point is due to our maintaining a ν^2 dependence in the model emissivity at long wavelength rather than the $\nu^{1.75}$ dependence discussed above.

4) Silicate and ice resonances in the near infrared

One additional application of this radiative transfer model is to the resonances at 10μ and 3μ which are seen in compact sources such as W3 IRS 5 and the BN object. These absorptions are commonly ascribed to silicate and ice materials in grains located in clouds along the line-of-sight to the source. Such an interpretation may be referred to as a cold-screen model with the absorbing grains extrinsic to the compact source. However, in the event that the same types of silicate and ice grains exist in the source itself, we have found that a substantial part of the absorption feature can occur as a radiative transfer effect in the object itself. This "intrinsic" model is capable of explaining all the absorption in BN and most of it in W3 IRS 5 (Kwan and Scoville 1976).

In an optically thick, distributed dust cloud (i.e., no sharp outer boundary) the effect arises as follows: Because the emissivity at the 10μ silicate resonance is greatly different from outside the resonance (we assumed a factor of six enhancement), the emergent flux will, in general, be different from that outside the resonance. For an optically thick cloud, we see into different regions of the cloud at different wavelengths, depending on the opacity at each. The higher opacity at 10μ means that the observed emission at this wavelength will arise in a region further out than where the emission just outside the resonance arises, and in general the dust temperature decreases with increasing radius because the energy is supplied from within. Then, whether the emission at 10μ is more or less than the emission outside the resonance depends on the balance between advantages for emission of a large surface area and a higher emissivity on the one hand against the disadvantage of a cooler temperature on the other. To evaluate these two, we first constructed a distributed dust model for BN which had parameters adjusted so that the calculated

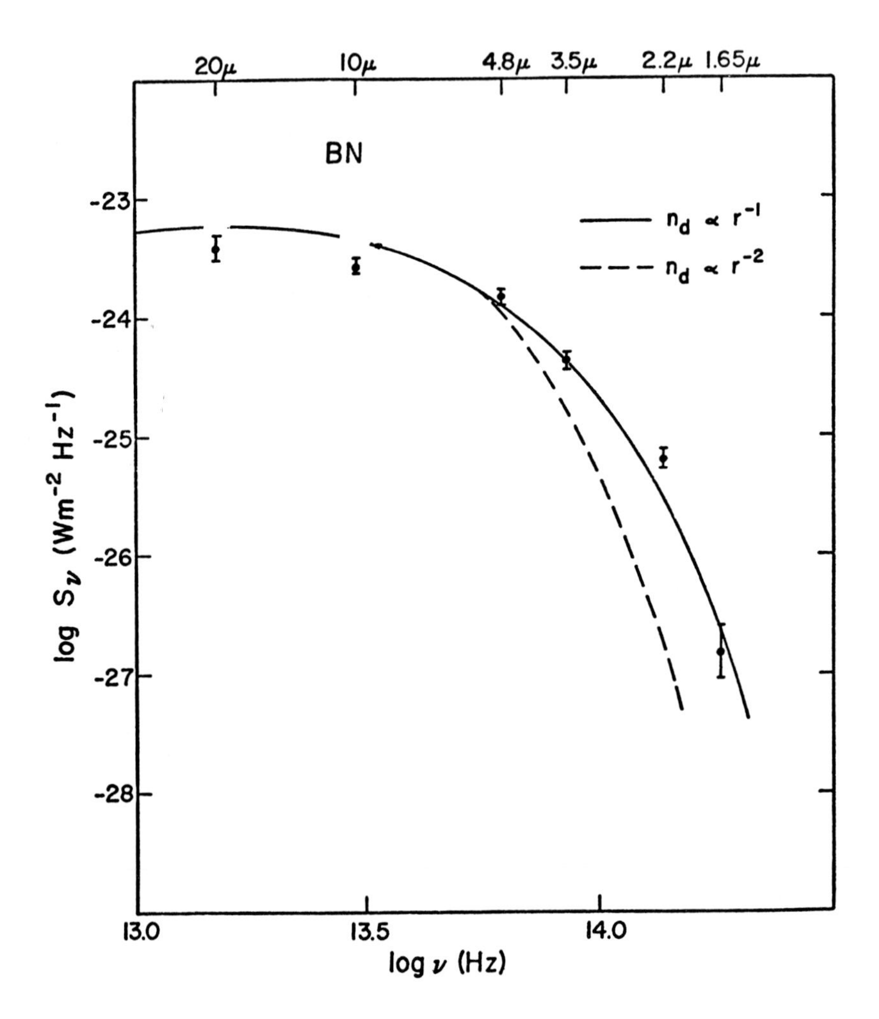

Figure 4: Continuum energy distributions of BN, calculated for both a r^{-1} (solid curve) and r^{-2} (dashed curve) density distribution. Observed data points are from Becklin, Neugebauer, and Wynn-Williams (1973). The emitted flux at 10μ and 20μ depend on the emissivities at the two silicate resonances and are discussed separately.

continuum spectrum adequately matched the observations. This model for which $L_* = 1.5 \cdot 10^3\ L_\odot$ and the optical depth at 5μ from the surface down to $r = 3 \cdot 10^{14}$ cm is unity is shown in Figure 4. In this same model we find that both the r^{-1} and r^{-2} density distributions give a silicate residual intensity only 30% of the neighboring continuum. The observed ratio is 25% (Gillett and Forrest 1973). Two quite definite predictions of this interpretation of the silicate feature as compared with extrinsic absorption models are that there should be a strong correlation between sources with low color temperature and deep silicate absorption and secondly there should be little or no 20μ absorption (also from silicates) for sources with color temperature greater than 200°K if the 10μ feature is intrinsic.

5) Relationship to HII regions and molecular clouds

A long standing issue in interpretation of far-infrared observations relates to the location of the dust grains emitting at $\lambda > 20\mu$. That is - are these grains situated within HII regions, in neutral gas just outside the HII regions, or in an entirely separate region of the molecular cloud where the grain heating must be from dust-embedded stars or protostars? The first situation has been reviewed here by Panagia, the second possibility has been discussed by Wright (1973). The Kleinmann-Low nebula on which we have based the foregoing discussion is quite clearly an example of the last possibility. Most persuasive is the occurrence of peak molecular line emission precisely at the position of the Kleinmann-Low nebula and the absence of any analogous association of KL with the isophotes of free-free emission from M42. A similar, good correlation between far infrared emission and molecular lines occurs in the NGC 6334 region. Figure 5 shows a map of the ^{12}CO emission from this region recently obtained by W. Irvine, R. Predmore, P. Wannier and myself. In particular, the luminous northern source observed by Emerson, Jennings, and Moorwood (1973) has no counterpart in the radio continuum maps, but instead is associated with an intense, extended cloud of molecular gas as evidenced by the CO emission. NGC 6334 is a region where I hope future detailed studies of infrared and molecular emission can help pin down the interrelationship.

Thermal coupling between the radiatively heated dust grains and ambient molecular gas (H_2) which could explain a correlation of far infrared and CO emission has been investigated by Goldreich and Kwan (1974). They indicated that the gas will approach thermal equilibrium with the dust ($T_K \to T_d$) via collisions of H_2 with grains (at $n_{H_2} > 10^4$ cm^{-3}). Unfortunately, even at $n_{H_2} = 10^4$ cm^{-3} the collisional rate is sufficient to give only $T_K \approx T_d/2$. Observational evidence, however, suggests that the coupling may be stronger. Over large regions in the galactic center the brightness temperature of $J = 1 \to 0$ CO emission (commonly interpreted as a measure of T_K) is within observational uncertainties equal to the

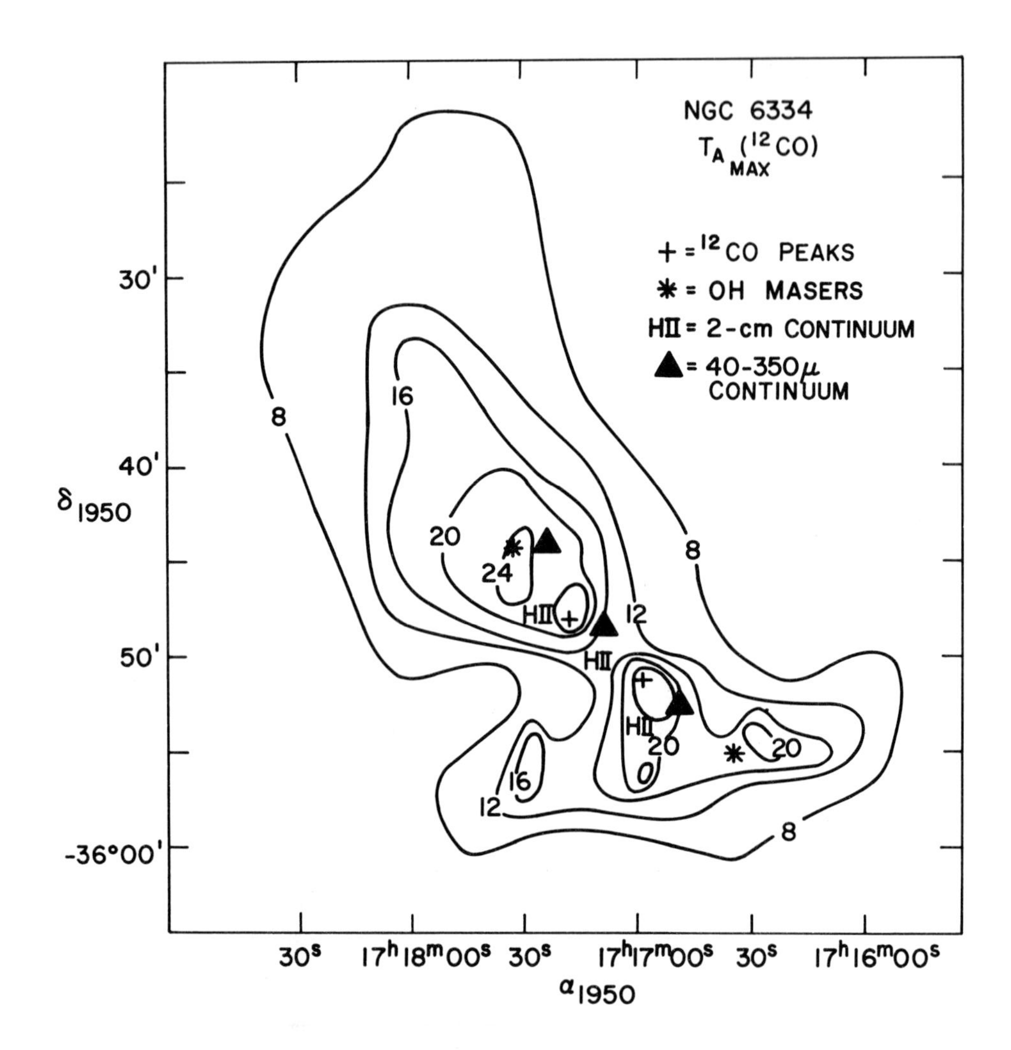

Figure 5: The peak antenna temperature T_A of the ^{12}CO emission in NGC 6334 is plotted against the locations of the far infrared peaks observed by Emerson et. al. (1973), the HII regions mapped by Schraml and Mezger (1969), and OH maser positions. The CO observations (HPBW = 2') were recently obtained with W. Irvine, R. Predmore, and P. Wannier at the University of Texas.

brightness temperature of 100μ infrared emission (Scoville et al. 1974, Hoffman et al. 1971).

A more rapid transfer of energy from dust to gas may exist through absorption of dust radiation in the rotational transitions of water molecules followed with collisional de-excitation by H_2. With numerical calculations of the H_2O excitation equilibrium, we have verified that this process is generally faster than direct H_2

collisions with grains. For the radiation field in a dust cloud characterized by $T_d = 45°K$ and $\tau_{100\mu} = 1$ and a density $n_{H_2} = 5 \cdot 10^3$ cm^{-3}, about 10% of the H_2O molecules will be excited into the 2_{12} level at an energy corresponding to 80°K (a $\lambda = 180\mu$ transition) above the ground state. From this level efficient heating may occur by absorption of a 75μ photon (transition from 2_{12} up to 3_{21}) and the subsequent collisional de-excitation. For a minimal abundance ratio $H_2O/H_2 = 3 \cdot 10^{-6}$ the rate of heating is 10^{-23} ergs sec^{-1} cm^{-3} at $T_K = 30°K$. [For comparison, the cooling rate by CO molecules at an abundance 10^{-4} n_{H_2} will be an order of magnitude less.] When T_K is raised to 40°K, we find that the situation is reversed and the H_2O cools the H_2. For these illustrative parameters the kinetic temperature will be 35°K in thermal equilibrium when the H_2O neither heats nor cools the H_2.

We, therefore, expect that even in clouds of moderate density with $n_{H_2} = 10^4$ cm^{-3} the gas temperature will be coupled to the dust brightness temperature. This will be true unless there is an additional heating input to the gas serving to raise T_K further.

Acknowledgements

This is contribution No. 233 of the Five College Observatories. The research of N. Z. Scoville is supported in part by NSF Grant MPS 73 - 04949-A03.

References

Becklin, E.E., Neugebauer, G., and Wynn-Williams, C.G. 1973, Ap. J. (Letters), 182, L7.
Emerson, J.P., Jennings, R.E., and Moorwood, A.F.M. 1973, Ap. J., 184, 401.
Gezari, D.Y., Joyce, R.R., Righini, G., and Simon, M. 1974, Ap. J. (Letters), 191, L33.
Gillett, F.C., and Forrest, W.J. 1973, Ap. J., 179, 483.
Goldreich, P. and Kwan J. 1974, Ap. J., 189, 441.
Harper, D.A., Low, F.J., Rieke, G.H., and Armstrong, K.R. 1972, Ap. J. (Letters), 177, L21.
Hoffman, W.F., Frederick, C.L., and Emergy, R.J. 1971, Ap. J. (Letters), 164, L23.
Irvine, W.M. and Pollack, J.B. 1968, Icarus, 8, 324.
Knacke, R.F. and Thomson, R.K. 1973, P.A.S.P., 85, 341.
Kwan, J. and Scoville, N.Z. 1976, Ap. J. (in press).
Rieke, G.H., Low, F.J., and Kleinmann, D.E. 1973, Ap. J. (Letters), 186, L7.
Schraml, J. and Mezger, P.G. 1969, Ap. J., 156, 269.
Scoville, N.Z. and Kwan, J.Y. 1976, Ap. J., June 15.

Scoville, N.Z., Solomon, P., and Jefferts, K.B. 1974, Ap. J. (Letters), 187, L63.
Werner, M.W., Gatley, I., Harper, D.A., Becklin, E.E., Loewenstein, R.F., Telesco, C.M., and Thomson, H.A. 1976, Ap. J. (in press).
Wright, E.L. 1973, Ap. J., 185, 569.

HIGH RESOLUTION MAPPING OF THE ORION NEBULA REGION AT 30, 50 AND 100 MICRONS

M. W. Werner, E. E. Becklin, I. Gatley, and G. Neugebauer

California Institute of Technology, Pasadena, California (USA)

ABSTRACT

The central $\sim 4' \times 4'$ region of the Orion Nebula-OMC1 complex has been mapped with 20" resolution in well-defined bands at 30, 50, and 100 μ from the NASA C-141 infrared observatory. These data provide detailed information about the distribution of matter, luminosity, and temperature in this region. The principal results are the following: A) At all three wavelengths, the surface brightness peaks sharply at the position of the infrared cluster, which includes the Becklin-Neugebauer and Kleinmann-Low objects, to the northwest of the Trapezium. The total 25 to 130 μ luminosity from a 30" region centered on this peak is $4 \times 10^4\ L_\odot$. B) This emission peak is resolved spatially at all three wavelengths, with a characteristic half-width of 35". C) The color temperature of the emission at the peak is ~ 120 K. The temperature decreases uniformly away from this peak as expected for a dust cloud heated by a central luminosity source. D) At all three wavelengths, the surface brightness of the emission in the direction of the H II region is measured to be less than 10% of the peak surface brightness. The color temperature of this emission is $\gtrsim 200$ K. E) At 50 μ and 100 μ, there is a strong suggestion of a ring of emission surrounding the H II region, which is well-correlated with the position of optically visible ionization fronts. F) The previously identified bar of emission associated with the ionization front near θ^2A Orionis is not resolved at any wavelength and is smaller than 30" in NW-SE extent. G) The total 25-130 μ emission from the region mapped is $\sim 2 \times 10\ L_\odot$. The infrared cluster and the Trapezium cluster appear to make roughly equal contribution to this luminosity.

G. G. Fazio (ed.), Infrared and Submillimeter Astronomy, 89.

HIGH RESOLUTION FAR-INFRARED MAPS OF HII REGIONS AND THE GALACTIC CENTER

E. L. Wright, G. G. Fazio, & D. E. Kleinmann
Center for Astrophysics
Harvard College Observatory & Smithsonian Astrophysical Observatory, Cambridge, Mass. (USA)

F. J. Low
Lunar & Planetary Laboratory, University of Arizona, Tucson, Arizona (USA)

ABSTRACT

The Center for Astrophysics - University of Arizona 102-cm balloon-borne telescope was used to produce high resolution (1') maps of the HII regions M17, M20, NGC 7538, and the Galactic Center. These maps will be presented and their relationship to ground-based infrared and radio observations will be discussed.

G. G. Fazio (ed.), Infrared and Submillimeter Astronomy, 90.

NEW HIGH SPECTRAL RESOLUTION AIRBORNE LINE OBSERVATIONS IN THE FAR INFRARED

J.P. Baluteau and E. Bussoletti*, N. Coron**
M. Anderegg, J.E. Beckman, H. Hippelein, and
A.F.M. Moorwood†

ABSTRACT

A high resolution ($\lambda/\Delta\lambda \sim 10^4$ Michelson interferometer has been used on NASA's AIRO (C-141) telescope in November-December 1975 to observe far infrared emission lines from HII regions. The instrument operates in the rapid scanning mode under computer control and high resolution spectra are computed, averaged and displayed on-line. Observations were made in the range 17.5-20μm and 70-100 μm, with spectral resolutions of 0.03 and 0.02 cm^{-1} respectively, on the Orion Nebula and Jupiter. Emission spectra of the stratosphere were also obtained in these wavelength ranges. Emission from SIII at 18.7μ has been detected in the Orion Nebula. The intensity of this line is discussed compared with theoretical predictions. Preliminary results concerning the observations of OIII (88μ) in the Orion Nebula are also presented.

*Group Infrarouge Spatial, Observatoire de Meudon, France

**LPSP, Verrières le Buisson, France

†Astronomy Division, ESTEC, Noordwijk, Netherlands

(Permanent address for J.P. Baluteau: Insituto di Fisica, Università di Lecce, Italy)

G. G. Fazio (ed.), Infrared and Submillimeter Astronomy, 91.

LAMELLAR GRATING INTERFEROMETRY AT 50μ TO BEYOND 600μ FROM THE GERARD P. KUIPER AIRBORNE OBSERVATORY

J. G. Duthie, J. L. Pipher, and M. P. Savedoff

University of Rochester, Rochester, New York (USA)

ABSTRACT

A lamellar grating interferometer of unique construction has been designed and successfully used on the Gerard P. Kuiper Airborne Observatory. The grating is spherical with the center of curvature in the prime focal plane of the telescope. This feature eliminates the need for collimating optics and thus improves the throughput of the system. Used in conjunction with a gallium doped germanium bolometer, cooled to liquid helium temperatures, the system is sensitive from 50μ to beyond 600μ with a present maximum resolving power $\frac{\lambda}{\Delta\lambda}$ of 500 at 100μ and the system NEP was 2×10^{-12} watts/$H_Z{}^{1/2}$. In flight the telescope operated open port ensuring minimum amount of attenuation. Observations were made with 20μ precipitable water above the aircraft. Interferograms were obtained of Jupiter at a resolution of $5cm^{-1}$ and of Mars, Saturn and the Kleinmann-Low region in Orion at somewhat lower resolution. Results of these preliminary observations from our first flights will be presented. This project was supported by NASA grant NGR 33-019-127.

G. G. Fazio (ed.), Infrared and Submillimeter Astronomy, 92.

FAR INFRARED POLARIZATION OF DUST CLOUDS

Brian Dennison

Center for Radiophysics and Space Research
Cornell University
Ithaca, New York 14853

ABSTRACT

The recently observed 10μ polarization of the Orion Nebula and the Galactic Center suggests that far infrared polarization may be found in these objects. Estimates are made of the degree of far infrared polarization that may exist in the Orion Nebula. The effects of radiation transfer are briefly considered and it is concluded that polarization observations over a range of wavelengths will be useful in deducing the optical depth structure of dust clouds.

G. G. Fazio (ed.), Infrared and Submillimeter Astronomy, 93.

PART III

S O L A R S Y S T E M

INFRARED OBSERVATIONS OF THE SUN

Pierre J.LÉNA
Université Paris VII et Observatoire de Meudon
92190 MEUDON, FRANCE

ABSTRACT

Infrared solar radiation is, almost in every case, formed in thermal equilibrium and provides true kinetic temperatures. Absolute measurements now cover the range 1-1000μ with improved accuracy. Center-to-limb variations at a number of wavelengths have been obtained on the ground, with balloons or aircrafts and contribute to improved models up to the chromosphere. Progress in imaging techniques has allowed the mapping of fine structures and active regions. The spectrum of the quiet Sun is mostly characterized by molecular lines, extremely temperature sensitive and probing the temperature and abundance fluctuations in granulation or active regions. The structure of the corona can be probed with fine structure lines. The near F corona composition has been studied and the nature and concentration of dust is better known.

1. INTRODUCTION

The infrared spectrum of the Sun, broadly defined to cover the spectral range 1μ-1mm, does not represent an observational problem which can only be solved by Space Research. Because of the progressive variation of atmospheric extinction with altitude (e.g. TRAUB, 1976) more information is slowly gained with increasing height, and ground based observations ($\lambda<25\mu$ $\lambda>300\mu$) are completed with airborne or balloon borne results in the intermediate spectral range.

Despite some gaps, the monochromatic flux of the integrated Sun is reasonably known with an average accuracy of $\pm$200K

G. G. Fazio (ed.), Infrared and Submillimeter Astronomy, 97-106.

over the whole IR, a situation which strongly contrasts with the situation only ten years ago, but which remains to be improved.

This is not true for the other properties of the Sun : the brightness distribution on the surface of the quiet Sun, even at a one-arc-minute scale, is not known accurately. The temporal variations of the surface brightness, connected to the chromospheric oscillations, begin only to be investigated. The contribution of the infrared range to the models of active regions or sunspots is quite non-existent; the energy spectrum of sudden events, such as bursts or flares, is only beginning to be extended and effects on the possible partial polarization of the radiation are investigated.

The radiation emitted on the disc in this wavelength range covers the deepest observable part of the photosphere at 1.6μ ($\tau_{5000} \sim 3$), minimum of the absorption coefficient of the H^-ion, up to the chromospheric level of the temperature minimum ($\tau_{5000} \sim 10^{-6}$), as observed in the millimetric range. Because of the known Local Thermodynamic Equilibrium of the main absorber, the H^-ion, the source-function accurately reflects the local kinetic temperature; on the other hand, the Rayleigh-Jeans situation which prevails in the whole IR equally weights the hottest and the coldest areas of the inhomogeneous surface, and the mean flux really represents an average local temperature, at least at the disc center.

The spectrum of the disc is mostly characterized by the contribution of the molecules, especially CO, which is both abundant and stable at high temperatures. Although this spectrum is not rich in lines, their likely formation in LTE and their sensitivity to temperature through the molecule dissociation, provides an accurate control of various models, and good estimation of abundances. Some studies of high-n emission or absorption lines (Rydberg lines) have also been undertaken.

The solar corona is optically extremely thin on the disc at all wavelengths in the IR, but may become significantly thicker and free from photospheric radiations when seen edgewise : this is especially true for two interesting contributions : the line emission of fine-structure levels of ionized atoms, such as Fe V, O IV etc. which is predicted to occur in the whole IR, and the continuum emission of the dust in the F-corona, which connects progressively with the interplanetary medium. A good review of the infrared solar spectrum has been given recently (DE JAGER,1974) to which we refer for detailed discussion of the continuous spectrum mostly. Another convenient review may be found in VERNAZZA et al. (1976).

2. THE MEAN CONTINUOUS SPECTRUM

The determination of a mean physical model of the photosphere and the chromosphere involves the contribution of numerous observations in the visible, ultraviolet and infrared, leading to a distribution of kinetic temperature T (z), density N (z), electron density N_e (z), and higher moments such as turbulent velocities ..... Current models are the Harvard Reference Smithsonian Atmosphere (Gingerich et al. 1971) or more recent improvements by Vernazza et al. (1973). The model gives the variation of temperature $T(\tau_\lambda = 1)$ and geometrical depth $Z(\tau_\lambda = 1)$ with λ. This is computed assuming a known absorption coefficient due to free-free transitions of H^-ion at $\lambda < 100\mu$ and including the free-free proton absorption and other contributions at longer wavelength. A monotonic increase of depth with decreasing wavelength is the consequence, with the almost exact relation : $\tau_{0.5} = (1.6/\tau_\lambda)^2$ (λ in microns). The available measurements are the monochromatic brightness at the disc center $B_\lambda(\mu=1)$ the overall solar flux Φ_λ and the relative center-to-limb variation $R_\lambda(\mu) = B_\lambda(\mu)/B_\lambda(1)$ with $\mu = \cos\theta$.

Because of the limited spatial resolution of many infrared measurements, the measured value $\emptyset_\lambda$ is often used as a good estimate of the quantity $B_\lambda(\mu = 1)$, because of the slow variation of the quantity $R_\lambda(\mu)$ over the disc surface, reflecting the weak gradients of $T(\tau_\lambda = 1)$ vs. λ. The absolute values of $\emptyset_\lambda$ are then converted in brightness temperature. Table 1 summarizes recent absolute measurements. The experimental difficulties are well reflected by the magnitude of the error bars and their rather slow decrease with time. A discrepancy remains at 50-60μ where only one set of measurements exists and needs confirmation.

TABLE 1
Recent Absolute Measurements of Solar Brightness Temperature

λ (microns)	T_B(extreme)	ΔT_B	Conditions	Authors
100-400	4200-4800K	$\pm$400K	Balloon	STETTLER et al. 1972
200 350 600	4100 4600 5200	$\pm$200K) $\mp$100K) $\mp$450K)	Balloon	d° 1975
300 to 1200	4500 6000	$\mp$300K) $\pm$250K)	Ground	DALL'OGLIO et al. 1974

A complete review of previous measurements is given by DE JAGER (1975) or by VERNAZZA (1976). Figure 1 gives the overall view of present measurements.

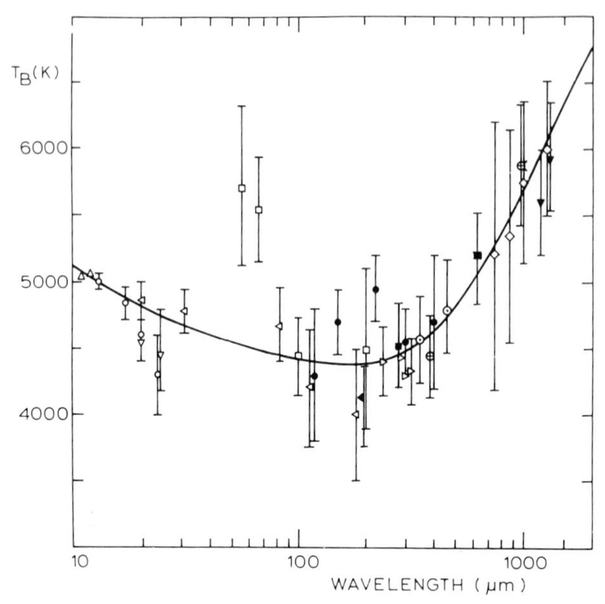

A general view of the Solar brightness temperature, taken in DE JAGER (1975) with more recent values (■,◀) from table 1.

The measurement of the center-to-limb variation, assumed to have circular symmetry, gives information on the quantity $dT_B/d\,\tau_\lambda$ and provides an independent check of the model. The main difficulties here are due to the fact that the significant variation occurs only at the very limb ($\mu \lesssim 0.1$) where : i) the diffraction limitations are severe, often force one to do eclipse observations at long wavelength; ii) the severe atmospheric fluctuations force one to do differential sky chopping, measuring $dR_\lambda\ (\mu)/d(\sin\theta)$ iii) the contribution of inhomogeneous chromospheric structures, such as spicules, seen edgewise, becomes significant and bias the measured mean temperature towards a poor representation of the true surface average.

Table 2 shows the range of values of (μ, λ) explored by various authors. The results seem fairly well established below 25 μ, where inhomogeneities would occur only for $\mu < 0.01$ and would not be measureable. The advent of good infrared spectrographs ensures a clean measurement of the continuum between spectral lines. Results become more debatable at longer wavelengths, where the contribution of active regions (on the disc) and optical rugosity effects at the limb make them more difficult to interpret. LINDSEY and HUDSON (1975) evaluated the amount of limb brightening expected in the submillimetric range on the basis of a purely spherically symmetric chromosphere, and showed the discrepancy with the observations : the observed brightening for $\mu>0.1$ is smaller than expected.BECKMAN et al. (1975) provides a measurement of the bright spike observed at the limb during an eclipse. Both sets of measurements can be explained with chromospheric inhomogeneities extending over more than a scale height, which are likely to be the spicules or their "roots" at shorter wavelengths (BECKMAN and ROSS 1976). Millimetric measurements are beyond the scope of this review but are also referenced in Table 2 and show similar features.

TABLE 2

Center-to-limb variation of Solar brightness

λ (microns)	μ	conditions	authors
0.7-2.5	1.0-0.03	Ground	PIERCE et al (1976)
2.5-10	1.0-0.05	"	ALLEN (1976)
10.0,18,22	1.0-0.06	"	LENA (1970)
24.3	1.0-0.09	Ground-Eclipse	NOYES et al (1968)
11-52	1.0-0.2	Balloon	MANKIN,STRONG (1968)
350-1000	1.0-0.2	Ground	LINDSEY, HUDSON (1976)
400-1200	0.20 arc sec	Aircraft	BECKMAN, ROSS (1976)
	from limb	Eclipse	BECKMAN et al. (1975)
800-1600	65 arc sec	Ground	ADE et al. (1974)[1]
3.3mm	1.0-0.1	Ground-Eclipse	SHIMABUKORO et al. (1975)
3mm-8mm	-	" "	HAGEN, SWANSON (1975)

(1) See discussion by KUNDU, SOUYAN (1975)

3. FINE STRUCTURES ON THE SOLAR SURFACE

3.1. Quiet Sun

The physically important structures are the granulation and supergranulation patterns, which both need a high spatial resolution (below 20 arc sec) and a good temperature discrimination : flux variations are of the order of a few percent. Although accurate temperature determination can help the construction of good models, it is difficult to show evidence of the presence of the patterns as they are seen in the UV or in the wings of visible lines.

TURON and LÉNA (1970) have obtained 10μ high resolution images of the solar surface, which show the low frequency tail of the granulation at 1-2 arc sec sizes.

To determine the temperature structure of the supergranules boundaries and its height dependence, WORDEN (1975) has investigated the frequency content of images, made at 1.72, 1.64 and 1.17μ. He finds temperature structures distributed on a supergranular size scale, with temperature differences ranging from

$\pm$ 200K at chromospheric levels, to -50K at deep photospheric levels, and tentatively assigned to magnetic field effects. His spatial resolution of 3" would be difficult to match at longer wavelengths.

The 300s oscillations have been mostly observed as velocity oscillations of 80$\pm$ 20m/s in photospheric spectral line profiles. Since they are not purely isothermal, they also appear as temperature fluctuations, and the comparison of phases between temperature and velocity can give indication on the propagation character of the waves. In the photosphere, the radiative relaxation time is of the order of 1s and the wave is isothermal. This time increases with height and in the upper photosphere, HUDSON and LINDSEY (1974) have convincingly observed temperature fluctuations at the 300-s period, with an amplitude of 3.0K rms. This work has been repeated and extended by LINDSEY (1976) and results are summarized in Table 3. The main conclusions are toward a rather short radiative relaxation time compared to $(300/2\pi)$s, and toward evanescent modes of the waves.

Higher spatial resolution is needed at submillimetric wavelengths to study the region of the temperature minimum, which has also been investigated by NOYES and HALL (1972) using CO vibration-rotation lines.

TABLE 3

Solar Atmosphere Oscillation in infrared continuum
(from LINDSEY 1976)

λ (μ)	Z(1)	ΔT (r.m.s.)K(2)	Aperture
10	150	3.0	10 arc sec
13	180	2.2	"
18	220	4.2	"
26	260	4.5	"
500	640	-	-
680		no measurable peak at 300s	2 arc min

(1) median of contribution function
(2) integrated in power spectrum from 2 to 4 mHz

3.2. Active Sun

We shall concentrate on recent interests toward non-thermal emission in the infrared and submillimetric, despite the fact that thermal radiation from active regions have been mapped in the infrared (LENA 1968, BECKMAN 1968). During a balloon flight,

DALL'OGLIO et al.(1972) have discovered some polarization of solar radiation beyond 100μ. Another ground measurement by DALL'OGLIO et al. (1974)confirmed this first result and showed a large degree of polarization, up to 12% at 600μ. The polarization is strongly wavelength dependent and the size of the "polarized sun" increases with wavelength attaining 60 arc min at 700μ. This high degree of polarisation could be explained by non thermal radiation and indicates that some control over flare events might need to be exerted when measuring absolute brightness in the far IR. Yet, the experiment has not been able to measure the instrumental polarisation on a source of similar spatial extent as the Sun.

The existence of a significant amount of radiation beyond the limb for $\lambda > 300\mu$, if permanent, should be noticed on eclipse results such as the ones of BECKMAN et al. (1975).

A general study of the physical mechanism creating infrared radiation in a flare event, and an evaluation of the flux levels, have been done by OHKI and HUDSON (1975). Main contributions would be synchrotron radiation, non-thermal and thermal Bremsstrahlung due to high ionisation of hydrogen. The contrast on the solar surface could reach a few percent.

A similar, although less extended, effort has been undertaken by CROOM (1976) who attempts to extend the radio burst classification in the submillimetric range.

HUDSON (1975) has investigated practical detection possibilities of temperature contrasts and finds a threshold of $\sim$5K at 20μ for ground based observations limited by atmospheric fluctuations, and of $\sim$30K at 350μ, limited by atmospheric extinction. These values would be somewhat reduced with space observations. This author has observed only one sub-flare, but his threshold values show that detectability of synchrotron radiation and white light continuum should be within reach in the present cycle of the Sun.

The contribution of the hot plasma can be separated from the synchrotron radiation by its polarisation, and an independent determination of electron temperature and magnetic fields could be obtained.

4. THE LINE SPECTRUM

If the line center optical depth corresponds to a layer below the temperature minimum, the line is in absorption; the reverse applies to lines formed above it.

The main infrared lines are the rotation-vibration lines of molecules, mostly CO in the photosphere. Many other molecules have been detected in the sunspot spectrum, and a basic reference for both spectra is the Atlas published by HALL (1973), which covers the spectral range 4000-8800 cm^{-1}. Beyond this, the reference work remains the Atlas published by MIGEOTTE et al. (1957), which covers only the photospheric spectrum, but up to 23.7μ.

Several balloon spectra cover parts of the fundamental vibration-rotation band of CO : MULLER, SAUVAL (1975), GOLDMAN et al. (1973). Comparison with predicted line profiles from current models basically confirms the LTE formation of CO lines. Some departures at the limb from the predicted brightening of the line core (NOYES, HALL 1972) can be better explained in term of increasing contribution of inhomogeneities seen edgewise, as mentioned above for the submillimetric brightening.

The isotopic lines also provide, especially in sunspot spectra, good determination of isotopic abundances : see e.g. the abundances of ^{17}O and ^{18}O as determined by HALL (1973).

There is no observation of the pure rotational spectrum of CO, although it should be measurable and would provide rotation temperatures which could be compared with vibration temperatures and confirm the LTE population of CO levels.

GREVE (1975) has studied the possible observation of high-n emission or absorption lines in the solar atmosphere, first pointed out by DUPREE (1968). He finds an observable range around 3 mm limited by the linewidth at one end, by the transition from emission to absorption at the other. The emission range below 200μ for low Z values or below for high Z values should also be observable and could provide interesting tool for the study of magnetic fields.

The emission of forbidden lines in the solar chromosphere and corona has been studied by DE BOER et al (1972), which predicts the wavelength and intensities of these lines when the transition region between the chromosphere and the corona is seen edgewise above the solar limb. Intensities comparable to the continuum with linewidth of the order of $\Delta\lambda/\lambda \sim 10^{-5}$ are predicted . Several of these lines have been observed in the near infrared, for example the Si X line at 1.43μ , the Mg VIII line at 3.03μ (MUNCH et al. 1967).

We also suggest to use these lines at longer wavelength (beyond 20μ) to study the coronal magnetic field, noting that the Zeeman separation increases much faster with wavelength than does the Doppler line width when the line is formed at temperatures of the order of 10^6K. Such effort would need the use of heterodyne radiometers, which were suggested by McELROY (1972) for the study of the solar continuum, and would be carried out in a suitable fashion on board Spacelab.

5. THE F. CORONA

Since the pioneering work of PETERSON (1963) and KAISER (1970), the infrared radiation of the F-corona has been carefully investigated and has given information on the density of particles near the Sun (1 to 20 $R_\odot$);for a review see e.g. LAMY (1974), on their concentration near the ecliptic plane, their composition and the existence of a vaporization zone (LENA et al. 1974).

Since the immediate solar neighbourhood offers physical conditions quite similar to conditions which may exist in HII regions, it seems to be an interesting laboratory to study the composition of dust, its behavior with respect to sputtering, its photoelectric charge, its sublimation, its dynamics .... etc, several questions which are difficult to answer in the interstellar medium.

6. CONCLUSION

Despite the fact that the infrared solar spectrum does not show a wealth of spectral lines comparable to the UV or XUV range, its scientific value remains significant because of the predominant situation of Local Thermodynamic Equilibrium which easily connects the measured radiation field and the local physical conditions. The progress of observations is fairly slow, because of the intrinsic difficulties of ground based observations and the limited number of space experiments (balloon or aircraft) which have been carried up to now. The very substantial progress in absolute accuracy of temperature determination within reach would help to assess the various forms of energy transfer in the transition zone between the photosphere and the chromosphere, both horizontally and vertically. The advent of a close solar maximum coïncides with developpements of sensitive detector systems and airborne platforms which were not available ten years ago and it may be expected that progresses will be achieved on active regions thermal and non-thermal emission.

REFERENCES

ADE,P.A., RATHER,J.D., CLEGG,P. Ap.J. 187,389, 1974
ALLEN,R. 1976 Thesis, in preparation
BECKMAN,J.E. Nature 220, 53, 1968
BECKMAN,J.E., ROSS,J., in "Far Infrared Astronomy" Pergamon 1976
BECKMAN,J.E., LESURF,J.C.G., ROSS,J. Nature, 254, 38, 1975
DE BOER,K.S., OLTHOF,H., POTTASH,S.R. Astr. Astroph.16,417 1972
CROOM,D.L. in Far Infrared Astronomy. Pergamon 1976
DALL'OGLIO,G., FONTI,S., GUIDI,I., MELCHIORRI,B., MELCHIORRI,F. NATALE,V., LOMBARDINI,P., TRIVERO,P. Infrared Physics. 14 327. 1974

DALL'OGLIO,G., GANDOLFI,E., MELCHIORRI,B., MELCHIORRI,F. NATALE,V. Infrared Physics. 13, 1, 1973
DUPREE,A.K. Ap. J. 152, L125 1968
GINGERICH,O., DE JAGER,C., 1968, Solar Physics 3, 4
GOLDMAN, MURCRAY, MURCRAY, WILLIAMS, Ap. J. 182, 581 1973
GREVE,A. Solar Phys. 40, 329 1975
HALL,D.N.B., An atlas of infrared spectra of the Solar photosphere and of sunspot umbrae. Kitt Peak Nat'l Observatory 1973
HALL,D.N.B., Ap J. 182, 977
HAGEN,J.P., SWANSON,P.N., Bull AAS. 7, 360, 1975
HUDSON,H.S., LINDSEY,C.A., Ap. J. 187, L 35 1974
HUDSON,H.S., Solar Physics 45, 69 1975
DE JAGER,C. Space Science Reviews 17, 645. 1975
KAISER,C.B., Ap. J. 159, 77 1970
KUNDU,M.R., SOU YAN,L. Sol. Phys. 44, 361, 1975
LAMY,P.L., Astron. Astrophys. 33, 191 1974
LENA,P., Solar Physics 7, 217, 1969
LENA,P.J., 1970 Astron. Astrophys. 4, 202
LENA,P., HALL,D., VIALA,Y., Astron. Astrophys. 37 81, 1974
LINDSEY,C.A., 1976, Preprint
LINDSEY,C., HUDSON,H.S., Ap. J. 1976. In press
MC ELROY,J.H. Appl Opt 11, 1619 1972
MANKIN,W., STRONG,J. Bull. AAS. 1, 200, 1969
MIGEOTTE,M. NEVEN,L., SWENSSON,J. Mem. Soc. Roy. Liège. 2, 1957
MULLER,C., SAUVAL,A. Astron. Astrophy. 39, 445 1975
MÜNCH,G., NEUGEBAUER,G. MC CAMMON,D. Ap.J. 149, 681 1967
NOYES,R.W., HALL,D.N. 1972, Ap. J. Lett. 176, L 89
NOYES,R.W., BECKERS,J.M., LOW,F.J. 1968 Solar Physics,3, 36
OHKI,K., HUDSON,H.S., Solar Physics 43 405, 1975
PETERSON,A.W., Ap J 138, 1218 1963
PIERCE,A.K., SLAUGHTER,C.D., WEINBERGER,D. 1976 Preprint
SHIMABUKORO,F., WILSON,W., MORI,T., SMITH,P. Sol. Phys. 40, 359, 1975
STETTLER,F.K., KNEUBÜHL,F.A., MULLER,E.A., 1972, Astr. Ap. 20,309
STETTLER,P., RAST,J., KNEUBÜHL,F.K. Solar Phys. 40, 337, 1975
TRAUB,W.A., In Far Infrared Astronomy. Pergamon 1976
TURON,P. LENA,P., Solar Phys. 14, 112, 1970
VERNAZZA,J.E., AVRETT,E.H., LOESER,R. 1973 Ap J. 184, 605
VERNAZZA,J.E., AVRETT,E.H.,LOESER,R. 1976 Ap J. Supp.Series 30, 1
WORDEN,S.P., Solar Phys. 45, 521, 1975

THE SOLAR BRIGHTNESS TEMPERATURE IN THE FAR INFRARED

E.A. Müller*, P. Stettler**, J. Rast**, F.K. Kneubühl**,
and D. Huguenin*

ABSTRACT

Measurements of the solar brightness temperature in the far infrared give information on the radial temperature distribution in the transition region between the photosphere and the chromosphere. Since the source of opacity in this wavelength range is due to the free-free absorption of the H^- ion, which is well understood, the absolute solar brightness measurements can readily be interpreted in terms of the temperature variation with height.

For the purpose of measuring the absolute solar brightness over a wide range of far-infrared and submillimeter wavelengths, a lamellar-grating interferometer was developed at the ETH Zürich. Incorporated in the Geneva Observatory gondola it was launched to an altitude of about 35 km from the balloon launching station of the CNES in Aire-sur-l'Adour, France. The tracking system of the Geneva Observatory gondola is described by D. Huguenin (1974) and the optical system of the balloon-borne lamellar grating interferometer is discussed in detail by Stettler et al. (1975). The experiment covering the wavelength range 200-600 μ revealed that at 200 μ the solar brightness temperature may reach the value of about 4100°K. Müller et al. (1975) mentioned that this result is not incompatible with the higher values of the temperature minimum derived from ultraviolet observations around λ1600Å. The emergent radiation in the 100-200 μ range originates at the same height as that of around λ1600Å. However in the far infrared the continuum source function S is equal to the Planck function B whereas in the ultraviolet S>B; thus the ultraviolet observations are expected to yield higher values of the temperature minimum than infrared measurements (Vernazza et al., 1976).

G. G. Fazio (ed.), Infrared and Submillimeter Astronomy, 107-108.

Recently an improved lamellar-grating interferometer was constructed and launched on September 19, 1975 to an altitude of 34.5 km. It recorded seven solar interferograms in the wavelength range of 80-250 μ. The resulting brightness temperature is constant between about 95 μ and 200 μ, thus revealing the flat temperature minimum of the photosphere/chromosphere transition region. Due to some calibration problems the temperature scale was adapted to the absolute calibration of previous flights covering the 200-600 μ range. Further absolute measurements are planned for the 80-250 μ wavelength range in order to establish precisely the value of the temperature minimum in the solar atmosphere.

Four of the seven spectra secured with the new lamellar-grating interferometer have a spectral resolution better than 0.5 cm^{-1}. This resolution allows the detection of a number of stratospheric molecular absorption lines (H_2O, O_2, O_3).

This research was supported by the "Fonda national suisse de la recherche scientifique."

References:

- Huguenin, D.: 1974, NASA TM X-62, 397, p. 167.
- Müller, E.A., Stettler, P., Rast, J., Kneubühl, F.K, and Huguenin, D.: 1975, Osservazione e Memorie Oss. di Arcetri No. 105, p. 90.
- Stettler, P., Rast, J., Kneubuhl, F. K., and Müller, E. A.: 1975, Solar Phys. 40, 337.
- Vernazza, J. E., Avrett, E. H., and Loeser, R.: 1976, Astrophys. J. Suppl. Ser. 30, 1.

* Observatoire de Geneve, 1290 Sauverny, Switzerland

** Solid State Physics Laboratory, ETH-Z Hönggerberg, 8049 Zürich, Switzerland.

INFRARED OBSERVATIONS OF THE PLANETS

G. H. Rieke
Lunar and Planetary Laboratory
University of Arizona
Tucson, Arizona 85721

Although infrared detectors were first trained on the planets many years ago, recent dramatic improvements in the sensitivity and sophistication of the instruments is resulting in surprising and fundamental discoveries. The inner planets and their satellites are in the process of a thorough exploration by spacecraft. However, it is the outer planets--so foreign from the earth--that hold the greatest fascination for many of us, and study of the infrared emission of these objects still depends largely on earthbased telescopes. Infrared observations are currently having a fundamental impact in a number of areas of outer solar system research. Four of these areas are discussed below--the nature of comets, the structure of the atmospheres of the outer planets, the surfaces of small bodies in the outer solar system, and the contraction history of the outer planets.

THE NATURE OF COMETS

Until recently, cometary observations concentrated on the gaseous component of the coma and tail, since it produces bright emission lines that can be studied spectroscopically. However, the thermal emission of the dust is dominant in the infrared, so much so that no gaseous emission lines have yet been detected there (Becklin and Westphal 1966; Westphal 1972). Infrared studies of comets, which only began in earnest a few years ago, have therefore brought an entirely new point of view to this subject. Interesting discoveries have been made regarding the structure of comet nuclei. Of wider interest, it has been possible to study the changes in a sample of solid material from the outer solar system as it is heated to high temperatures.

G. G. Fazio (ed.), Infrared and Submillimeter Astronomy, 109-120.

Good infrared observations of comets are now available at heliocentric distances ranging from 0.15 AU (equilibrium temperature ~720°K) to 2.1 AU (~190°K).

Comet Kohoutek (1973f) is by far the most thoroughly observed in the infrared. Major campaigns were carried out at the California Institute of Technology, University of Minnesota, and University of Arizona, and occasional measurements were made by other observers. These observations are summarized in Figure 1, where they have been corrected to give the spectra corresponding to a uniform distance of 1 AU from the earth and circular beam 10,000 km in diameter centered on the nucleus.

The color temperature of the dust grains increases dramatically near the sun and reaches a maximum of ~1000°K, somewhat higher than the equilibrium temperature for large, conducting particles at the appropriate heliocentric distance. The relatively high temperature of the grains implies that their infrared emissivity is smaller than their optical absorption efficiency, which is ~0.8 (Rieke and Lee 1974; Ney 1974a). There is some indication that the emissivity goes through a minimum short of 10μ (Rieke and Lee 1974), so the true grain temperature may be even slightly higher than the color temperature.

The silicate emission feature at 10μ is clearly seen in nearly all the observations, although not at heliocentric distances greater than 1.7 AU. Most measurement sets at 20μ were not sufficiently detailed to delineate the feature expected there, but the most complete measurements, made at a heliocentric distance of 0.37 AU approaching the sun, show an emission peak nearly as large as at 10μ. Higher-resolution spectra taken at about the same time show the 10μ feature to be without spectral structure (Merrill 1974). There appear to be fluctuations both in the strength of the 10μ feature and in the shorter-wavelength spectrum, but no systematic trends, other than those expected from the increased grain temperature near the sun.

It has been possible to confirm some of these conclusions from observations of other comets. The splitting of the nucleus of Comet West (1975n) allowed a more sensitive search for possible fluctuations in the spectrum. Changes on a time scale of days found in the strength of the 10μ feature at nucleus D could be compared with simultaneous measurements of nucleus A to establish their reality beyond reasonable doubt. Much higher quality observations than for Comet Kohoutek again showed a disappearance of the silicate emission feature near 2 AU (Rieke 1976). Measurements of Comet Bradfield (1974b) showed a sudden decrease in the strength of the silicate feature followed by an abrupt decrease in the infrared flux (Ney 1974a).

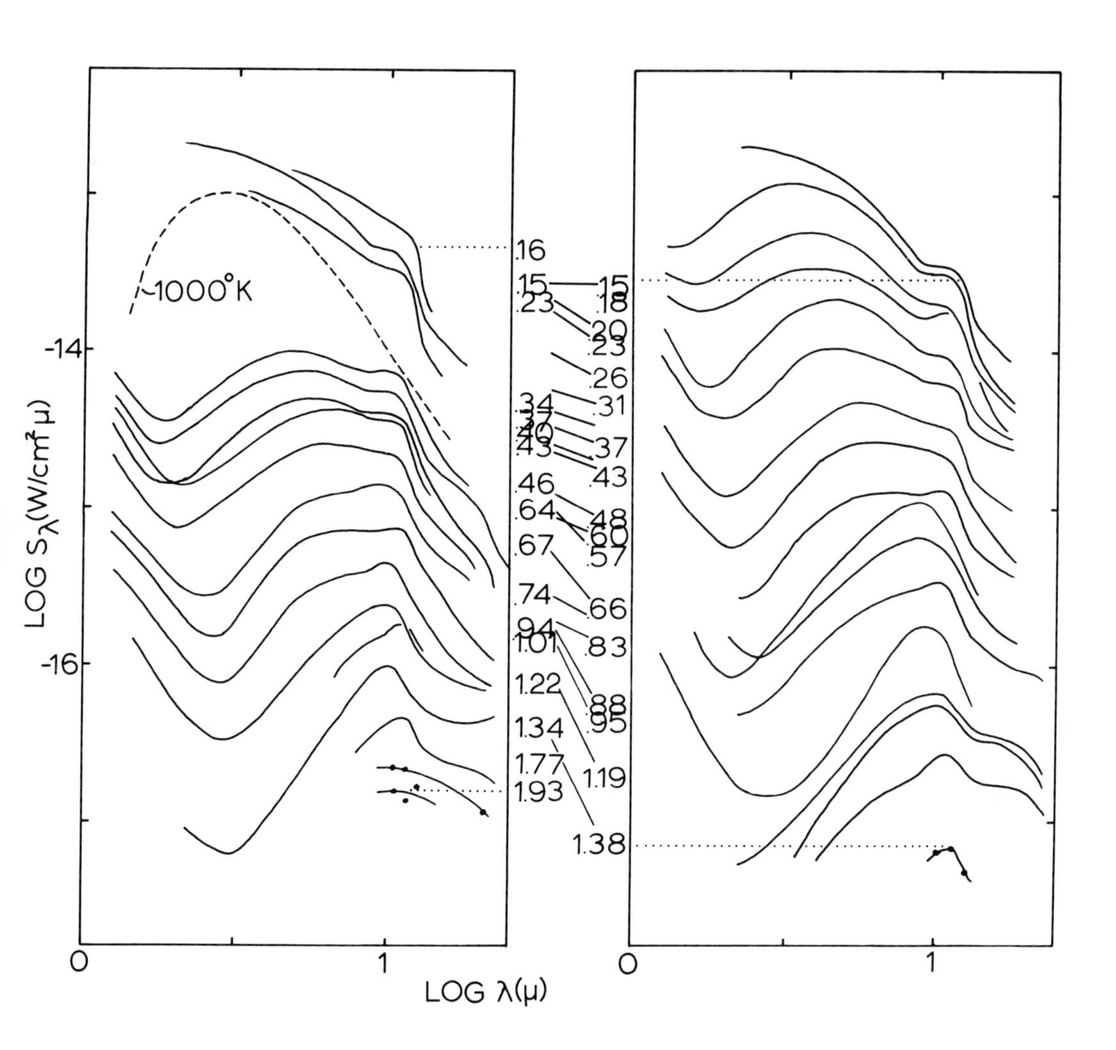

Figure 1. Infrared Spectra of Comet Kohoutek, from the observations of Rieke and Lee (1974), Ney (1974b), Gatley et al. (1974), Zeilik and Wright (1974), and Rieke et al. (1975). The spectra have been normalized to correspond to a 10,000 km region, 1 AU from the earth. The heliocentric distance in AU is labelled at the flux seen at 11μ. Spectra for the comet approaching the sun are on the left, and leaving the sun on the right.

In principle, the spectral changes could be caused by changes in grain composition or size. However, observations of the antitail of Comet Kohoutek by Ney (1974b) established a close connection with dynamical calculations of dust tail geometry (e.g. Finson and Probstein 1968; Sekanina 1974). These calculations show than an antitail can appear through the ejection of large particles, with diameters ~100μ. Ney found no silicate feature in the antitail, implying that the grains are so large that the emission band is saturated. A further connection between the dynamics of dust grains and the infrared observations is implied from Sekanina's (1976) argument that the broad, truly synchronic streamers in comet tails represent outbursts of dust production. These outbursts are probably associated with the rapid fluctuations seen in the infrared spectrum.

The fluctuations in the infrared spectrum occur too rapidly to be reconciled with an "onion skin," or layered model of the nucleus, since the erosion must be uneven and it is unlikely that a layer would be exhausted simultaneously over the whole surface. It is more likely that the material lies in pockets in a matrix of more resistant material, and that the infrared fluctuations and broad tail streamers are a consequence of the successive exhaustion of different pockets containing differing size distributions of silicate grains. In Comet Bradfield, we may have even seen the exhaustion of the last available pocket, leaving only the resistant core matrix.

The silicate grain material appears to be similar to that seen in the general interstellar medium, particularly with regard to its lack of spectral structure in the 10μ feature. Day (1974) has produced plausible analogues to this material by slowly growing amorphous silicates in the laboratory. A general property of these materials is that they crystallize and show spectral structure after being heated to 600-700°C. In most cases, the strength of the absorption coefficient also increases dramatically. A comet that passes close to the sun provides a unique test for the presence of such grains in a celestial object. The lack of any detectable change in the strength of the silicate feature in Comet Kohoutek, even at a heliocentric distance of 0.15 AU, already puts strong constraints on the nature of the grain material. Measurements at higher spectral resolution would provide a much more stringent test.

The disappearance of the silicate feature near heliocentric distances of 2 AU correlates well with the expected stability of water clathrate grains at this distance (Sekanina 1973). However, the albedo of the grains did not change significantly at this distance in Comet West, indicating that water ice may act as a glue that holds the grains in clumps until they are well removed from the nucleus.

THE STRUCTURE OF THE ATMOSPHERES OF THE OUTER PLANETS

Infrared studies of Jupiter have recently been reviewed by Ingersoll (1976) and Ridgway et al. (1976); it is not appropriate to cover the same ground less thoroughly here. However, a few areas of immediate future promise should be mentioned.

It is well-known that there is a window in the Jovian atmosphere at 5μ which permits observations at this wavelength to penetrate to the tops of the cloud layers, or to even deeper levels where there are holes in the clouds. Brightness temperatures as high as 300°K occur (Westphal 1969), and many of the brightest areas are blue in color (Keay et al. 1973), possibly indicating rayleigh scattering deep in the clear atmosphere. The surface brightnesses tend to cluster around three values (Armstrong et al. 1976; Terrile and Westphal 1976). The exact interpretation of these values is unclear--the smallest could correspond to reflected sunlight from areas of high albedo, thermal emission, or a combination, while the largest could be the top of a hot, deep cloud layer or the result of increasing atmospheric opacity at great depths. Nonetheless, at least two cloud layers appear to be necessary to explain the observations.

Water and phosphene absorption lines have been found in the 5μ window (Larson et al. 1975; Larson 1976). Application of higher spatial and spectral resolution should permit measurement of the pressure broadening of these lines in regions of differing surface brightness, thus obtaining detailed pressure-temperature-composition information about the atmosphere, and locating the cloud layers with regard to pressure and temperature.

Further characteristics of the Jovian cloud system can be deduced from coordinated spatial mapping at 5μ and other infrared wavelengths. A particularly interesting phenomenon is the large range of albedo at 2μ, with the highest albedo in the region of brightest 5μ sources and very low albedo where the 5μ surface brightness is lowest (Armstrong et al. 1976; Rieke and Armstrong 1976). The highest cloud layer is apparently very dark at 2μ.

The other outer planets have so far been subjected only to cursory exploration compared with Jupiter. Some of the most interesting discoveries concern the structures of their upper atmospheres.

Saturn exhibits limb brightening both in the 12μ ethane and 7.7μ methane bands (Gillett and Orten 1975; Gillett 1975). This behavior is similar to that of Jupiter (Gillett and Westphal 1973), and indicates a temperature inversion in the upper atmosphere. However, the limb brightening on Saturn occurs only over the south polar cap (Rieke 1975).

Even more peculiar are the behavior of Uranus and Neptune (Gillett and Rieke 1976). Uranus has a negligible upper atmospheric temperature inversion, judged from the complete absence of detectable methane and ethane emission features. In fact, the upper limit to the brightness temperature at 12μ is only slightly above the measured brightness temperature at 20μ. In contrast, bright emission is seen in the spectrum of Neptune at both 8 and 12μ.

In general, methane and ethane will produce temperature inversions in the upper atmosphere of an outer-solar-system body if they exist in a region where the pressure is low enough that the far-infrared pressure-induced opacity is low, but where their density is sufficiently high to absorb sunlight in near-infrared bands and produce sufficient opacity to emit thermally in the middle-infrared bands. The variety of properties exhibited by the upper atmospheres of Saturn, Uranus, and Neptune indicates radical differences in the atmospheric circulation of these bodies. No satisfactory explanation of these phenomena has been advanced.

SURFACES OF SMALL BODIES IN THE OUTER SOLAR SYSTEM

The surfaces of the satellites of the major planets are of interest both in themselves and because they provide constraints on theories for the contraction history of the central body. Thermal infrared observations can be used to measure the diameters of these objects, some of the surface characteristics, and in cases where eclipses occur, the thermal inertia of the uppermost layers. Because the strongest absorption bands of most ices are in the near-infrared, observations in that spectral region can often identify the dominant constituents of the surface. Discussion of these subjects can be found in recent reviews by Morrison and Cruikshank (1974) and Morrison (1976a).

Water ice has now been observed on Europa, Ganymede (Pilcher et al. 1972; Fink et al. 1973), the rings of Saturn (Pilcher et al. 1970; Kuiper et al. 1970), and four satellites of Saturn (Fink et al. 1976; Morrison et al. 1976). It is feasible to extend such observations to a number of other satellites at resolution adequate to identify the dominant surface constituents. A probable identification of methane ice on Pluto has been made on this basis (Morrison 1976b).

The temperature dependence of the shape of the ice absorption at 1.7μ has been used to show that the ice on Europa and Ganymede is at ∿100°K (Fink and Larson 1975). Particularly for Ganymede, this temperature is well below that determined radiometrically.

This difference arises because the radiometric measurements sense primarily the flux from dark, warm areas, while the spectroscopic measurement applies to a substance that can survive at the heliocentric distance of these satellites only because its high albedo allows it to remain at a low temperature (e.g., Lebofsky 1975).

The thermal spectrum of a satellite surface divided into low- and high-temperature areas should have increasing brightness temperature with decreasing wavelength as a result of the increasing dominance of the emission of the hot areas on the Wien tail of the blackbody curve. Broadband 10- and 20-μ measurements of the Galilean satellites indicate such a trend (Morrison and Cruikshank 1974), and its existence is now firmly established through narrow-band photometry from 8 through 13μ (Rieke, Howell, and Fink 1976). Consistent models can be constructed which account for the thermal spectrum, the visual albedo, and the extent of ice coverage of the surface. These models need assume only two kinds of surface material: ice, at T∿100°K; and a dark material at T∿160-170°K. Therefore, temperature variations over the surface probably play a stronger role in the thermal spectra of the Galilean satellites than do wavelength-dependent emissivity variations.

High-quality eclipse cooling curves were measured for Ganymede and Callisto by Morrison and Cruikshank (1973) and Hansen (1973). These curves were analyzed in terms of two-layer surface models in which a top layer of very low thermal inertia (finely divided dust) cooled rapidly, and a warm bottom layer of high thermal inertia accounted for the thermal emission far into eclipse. Stringent limits were placed on the amount of high-inertia material that could be exposed on the surface.

The realization that the surface is divided into warm areas and cold areas suggests an alternate explanation for the cooling curves. The upper limit to the amount of exposed high-inertia material applies only if the material is dark and therefore at the higher temperature. If the exposed ice has high inertia, it can account for the emission that was provided in the earlier models by the underlying surface layer. R. Howell (1976) has demonstrated that the eclipse data can be fit excellently with two-temperature models in which the low temperature regions have high thermal inertia.

A clear choice cannot yet be made between these two kinds of models. Since they represent two extreme possibilities, the truth may also lie in between. However, the recent discovery that the brightness temperature of Callisto at 3.3 mm is well below the value in the thermal infrared (Ulich and Conklin 1976) may be difficult to reconcile with two-layer models, since it implies that the temperature falls rapidly as one goes beneath the surface.

The observations therefore indicate that the surfaces of JII-IV consist of clean ice and powdery, dark dust. It is easy to understand how such surfaces would evolve. Since ice is only marginally stable at the heliocentric distance of these objects, ice whose surface was darkened appreciably would heat up and evaporate, leaving a residue of impurities. The dominant impurity may be finely divided, dark, silicate-rich dust, as is found in comets. The residue would further darken the surface, accelerating the evaporation of the underlying ice until a sufficiently thick surface layer had accumulated to insulate the underlying ice from further erosion. Only clean ice could avoid this process.

Io does not share the properties of the outer Galilean satellites. Its near-infrared albedo is high and featureless (Morrison and Cruikshank 1974) and its thermal emission has an anomalous peak near 8μ (Rieke, Howell, and Fink 1976). Although we are beginning to acquire some understanding of the surfaces of the other satellites, we are only now discovering the problems to be solved for Io. Many more observations are needed to define the peculiar properties of this satellite.

CONTRACTION HISTORY OF THE OUTER PLANETS

Infrared measurements can help put two kinds of constraint on theories of the contraction of the outer planets. A thorough understanding of the properties of the satellites is needed to determine the heat generation in the early stages of the formation of the planet. This information is particularly useful where a systematic trend of properties is observed, as is the case for the Galilean satellites where clear trends are seen in density, percentage of surface coverage by ice, and other properties. Recent progress in understanding these bodies has already been described. The heat generation in the later stages of contraction can be determined by infrared measurements of the excess emission above the absorbed solar energy.

The discovery of the internal energy of Jupiter (Aumann et al. 1969) has been followed by a number of measurements to refine estimates of this quantity (Ingersoll 1976; Harper 1976). The actual measurements are in good agreement. However, the measurements by Pioneer 10 and 11 sample less than 50% of the total output and must be extrapolated to estimate the internal energy. Earth-based measurements cover the entire spectrum and indicate a larger internal energy than is deduced from the spacecraft measurements. Despite small discrepancies, it is well-established that the internal energy source is about 1.2 times the absorbed solar flux, with an uncertainty determined at least as much by the uncertainty in the phase integral as by the calibration and interpretation of the infrared measurements.

Detailed calculations of the contraction of homogeneous, adiabatic models of Jupiter have been carried out by Graboske et al. (1975). Good agreement is found with the observed radius, and the predicted luminosity is less than a factor of two below the measurements. Thus, the agreement is already very good, and the remaining discrepancy may be removed by improvements in the theoretical model, the infrared measurements, and the value of the bolometric albedo. It is also possible that an additional energy source, such as chemical differentiation (Salpeter 1973), contributes to the emitted flux. The trend of properties observed in the Galilean satellites is in agreement with the thermal histories indicated by the calculated models of Jupiter.

The measurements of Saturn are not currently of the same quality as those of Jupiter. The phase integral can only be guessed from analogy with the measurements of Jupiter, while the low angular resolution of the far-infrared instruments has made it impossible to measure the disk separately from the rings. The situation in both areas will improve after spacecraft have visited the planet and when the opening angle of the rings as viewed from the earth has decreased so ring emission contributes negligibly to earth-based far infrared measurements.

Recently it has been possible to measure the disk radiation at high angular resolution out to 40μ (Rieke 1975). The total flux to this wavelength already accounts for 80±15% of the absorbed solar radiation, where the quoted error corresponds to the uncertainty in the phase integral. Since there is no cut-off in the spectrum beyond 40μ, the existence of an internal energy source in Saturn is well established. An estimate of the size of this source requires that the flux from the rings be subtracted from the far infrared measurements. A previous estimate (Rieke 1975) was based on far infrared measurements that appear to be systematically slightly high. Based on a recent recalibration of these measurements (Wright 1976) and an independent set of new observations (Harper 1976), Saturn emits ∿1.8 times as much energy as it absorbs from the sun. Major uncertainties in this parameter arise from the phase integral, the ring models, and the far infrared measurements.

Theoretical calculations of the contraction history of Saturn have been carried out by Pollack et al. (1976), using techniques and models similar to those employed for Jupiter by Graboske et al. (1975). In this case, the calculated radius is slightly too large. The predicted luminosity is somewhat smaller than measured, but in view of the uncertainties in the measurement this discrepancy need not be taken very seriously. In another article, Pollack et al. (1976) show how the calculated contraction history is consistent with the known properties of the rings and satellites. In particular, the methane-rich atmosphere of Titan correlates with their calculation that it is the innermost satellite where methane-containing ice could condense. The low densities and extensive surface ice on the other satellites and ring particles are also consistent with their predictions.

Recent measurements of Uranus (Fazio et al. 1976; Harper 1976) indicate that its internal energy, if any, is very small.

CONCLUSION

Infrared measurements have played an important role in the rapid expansion of our knowledge of the outer solar system. It is particularly gratifying to see the combination of different kinds of studies into a more comprehensive picture of complex phenomena. Examples are 1.) the relation of cometary dust to interstellar material and to the surface material on outer solar system satellites, and 2.) the connection through calculations of contraction history between the internal energies of the large planets and the properties of their satellites.

Much of our knowledge of the infrared properties of the outer planets is derived from groundbased observations, where high angular and spectral resolutions can be achieved relatively easily. Nontheless, many of the most important discoveries, such as the internal energy of Jupiter and the presence of water and phosphene deep in the Jovian atmosphere, have come through airborne observations. Obviously, since they are cold bodies, a fundamental understanding of the outer planets awaits the deployment of more sophisticated air- and space-borne instruments operating in the far infrared.

ACKNOWLEDGEMENTS

Helpful discussions and communication of results before publication by K.L. Day, F.C. Gillett, D.A. Harper, H.P. Larson, F.J. Low, D. Morrison, J.B. Pollack, and E.L. Wright are gratefully acknowledged. This work was supported by NASA.

REFERENCES

Armstrong, K.R., Minton, R.B., Rieke, G.H., and Low, F.J. 1976, Icarus, in press.
Aumann, H.H., Gillespie, C.M., and Low, F.J. 1969, Ap.J., 157, L69.
Becklin, E.E., and Westphal, J.A. 1966, Ap.J., 145, 446.
Day, K.L. 1974, Ap.J., 192, L15.
Fazio, G.G., Traub, W.A., Wright, E.L., Low, F.J., and Trafton, L. 1976, Ap.J. (Letters), in press.
Fink, U., Dekkers, N.H., and Larson, H.P. 1973, Ap.J., 179, L155.
Fink, U., and Larson, H.P. 1975, Icarus, 24, 411.
Fink, U., Larson, H.P., Gautier, T.N., and Treffers, R. 1976, Ap. J. (Letters), in press.
Finson, M.L., and Probstein, R.F. 1968, Ap.J., 154, 353.
Gatley, I., Becklin, E.E., Neugebauer, G., and Werner, M.W. 1974, Icarus, 23, 561.
Gillett, F.C. 1975, private communication.
Gilett, F.C., and Orten, G.S. 1975, Ap.J., 195, L47.
Gillett, F.C., and Rieke, G.H. 1976, in preparation.
Gillett, F.C., and Westphal, J.A. 1973, Ap.J., 179, L153.
Graboske, H.C., Pollack, J.B., Grossman, A.S., and Olness, R.J. 1975, Ap.J., 199, 265.
Hansen, O.L. 1973, Icarus, 18, 237.
Harper, D.A. 1976, private communication.
Howell, R. 1976, private communication.
Ingersoll, A.P. 1976, Sp. Sci. Reviews, 18, 603.
Keay, C.S.L., Low, F.J., Rieke, G.H., and Minton, R.B. 1973, Ap.J., 183, 1063.
Kuiper, G.P., Cruikshank, D.P., and Fink, U. 1970, Sky and Tel., 39, 80.
Larson, H.P. 1976, private communication.
Larson, H.P., Fink, U., Treffers, R., and Gautier, T.N. 1975, Ap.J., 197, L137.
Lebofsky, L.A. 1975, Icarus, 25, 205.
Merrill, K.M. 1974, Icarus, 23, 566.
Morrison, D. 1976a, in Planetary Satellites, ed. J.A. Burns, Univ. of Arizona Press, in press.
Morrison, D. 1976b, private communication.
Morrison, D., and Cruikshank, D.P. 1973, Icarus, 18, 224.
Morrison, D., and Cruikshank, D.P. 1974, Sp. Sci. Rev., 15, 641.
Morrison, D., Cruikshank, D.P., Pilcher, C.B., and Rieke, G.H. 1976, Ap.J. (Letters), in press.
Ney, E.P. 1974a, Icarus, 23, 551.
Ney, E.P. 1974b, Ap.J., 189, L141.
Pilcher, C.B., Chapman, C.R., Lebofsky, L.A., and Kieffer, H.H. 1970, Science, 167, 1372.
Pilcher, C.B., Ridgway, S.T., and McCord, T.B. 1972, Science, 178, 1087.

Pollack, J.B., Grossman, A.S., Moore, R., and Graboske, H.C. 1976, preprint.
Ridgway, S.T., Larson, H.P., and Fink, U. 1976, in Jupiter, ed. T. Gehrels, Univ. of Arizona press, in press.
Rieke, G.H. 1975, Icarus, 26, 37.
Rieke, G.H. 1976, in preparation.
Rieke, G.H., and Armstrong, K.P. 1976, in preparation.
Rieke, G.H., Howell, R., and Fink, U. 1976, in preparation.
Rieke, G.H., and Lee, T.A. 1974, Nature, 248, 737.
Rieke, G.H., Low, F.J., Lee, T.A., and Wisniewski, W. 1975, Comet Kohoutek, ed. G.A. Gary, NASA--U.S. Gov. Printing Office, p. 175.
Salpeter, E.E. 1973, Ap.J., 181, L83.
Sekanina, Z. 1973, Astrophys. Lett., 14, 175.
Sekanina, Z. 1974, Sky and Tel., 47, 374.
Sekanina, Z. 1976, Sky and Tel., 51, 386.
Terrile, R.J., and Westphal, J.A. 1976, Icarus, in press.
Ulich, B.L., and Conklin, E.K. 1976, Icarus, 27, 183.
Westphal, J.A. 1969, Ap.J., 157, L63.
Westphal, J.A. 1972, Comets--Scientific Data and Missions, ed. Kuiper and Roemer, Lunar and Planetary Laboratory, Univ. of Arizona, p. 23.
Wright, E.L. 1976, Ap.J. (Letters), in press.
Zeilik, M., and Wright, E.L. 1974, Icarus, 23, 577.

FAR INFRARED SPECTRAL OBSERVATIONS OF VENUS, MARS AND JUPITER

Dennis B. Ward, George E. Gull and Martin Harwit

Center for Radiophysics & Space Research
Cornell University, Ithaca, New York, U.S.A.

ABSTRACT

45 to 115 micron spectra of Venus, Mars and Jupiter have been obtained at a resolving power of $\sim$ 10, observing from the NASA Lear Jet at an altitude of 13.7 km. The results are calibrated with lunar observations, and show Mars and Venus to have relatively constant brightness temperatures over this wavelength region, with Mars appearing somewhat cooler beyond 80μ. The brightness temperature of Jupiter decreases toward longer wavelengths, in general agreement with the models of Orton (1975).

G. G. Fazio (ed.), Infrared and Submillimeter Astronomy, 121.

PART IV

GALACTIC CENTER, EXTRAGALACTIC SOURCES AND SUBMILLIMETER RADIATION

INFRARED RADIATION FROM THE GALACTIC CENTER

E.E. Becklin

California Institute of Technology, USA

ABSTRACT

A review of recent observations of the infrared radiation from the galactic center region is presented. Particular emphasis is placed on recent one arc minute resolution 30- to 150-μm maps made from a balloon-borne telescope and from the C 141 air-borne telescope. It is shown that most of the far infrared radiation from the central 10 pc of the galactic center comes from dust heated by the bright stellar distribution observed at 2.2 μm. The infrared radiation from other components, such as Sgr B_2 is also discussed.

G. G. Fazio (ed.), Infrared and Submillimeter Astronomy, 125. *All Rights Reserved.*

FAR INFRARED SPECTRAL OBSERVATIONS OF THE GALACTIC CENTER REGION FROM THE GERARD P. KUIPER AIRBORNE OBSERVATORY

L. J. Caroff, E. F. Erickson, D. Goorvitch,
J. D. Scargle, J. P. Simpson, and D. W. Strecker

NASA-Ames Research Center
Moffett Field, CA, U.S.A.

ABSTRACT

Low resolution spectra in the region 45-250 microns of SgrA and SgrB2 were obtained in July/August, 1975, with the 91 cm telescope on the Gerard P. Kuiper Airborne Observatory. The spectra were taken with a single-beam Michelson interferometer utilizing a Mylar beam-splitter and a Ge-Ga bolometer with a beam diameter of 1.4 min. FWHM. Preliminary results give mean brightness temperatures for SgrA of roughly 90°K and for SgrB2 of roughly 40°K. SgrA was observed with a resolution of approximately 6 cm^{-1}, while the resolution for B2 is approximately 3 cm^{-1}. In addition, the HII region NGS 7538 was also observed at approximately 9 cm^{-1} resolution. The data will be presented and discussed.

G. G. Fazio (ed.), Infrared and Submillimeter Astronomy, 126.

HIGH-RESOLUTION FAR INFRARED OBSERVATIONS OF SGR B2, W49, AND DR21-W75

P. M. Harvey, M. F. Campbell, W. F. Hoffmann

Steward Observatory, University of Arizona
Tucson, Arizona

ABSTRACT

High resolution (∿20") maps at 53μ and photometry at 53, 100, and 175μ are presented for the HII regions Sgr B2, W49, and DR21-W75. In general strong far infrared emission is coincident with HII condensations although several interesting exceptions are described. Simplified models are fit to the far infrared spectra to derive estimates of dust temperatures and optical depths. The models indicate that these sources have relatively large far infrared optical depths, ∿0.1 - 1 at 100μ, and temperatures of 30 - 50K. The amount of dust indicated by the far infrared emission is consistent with that expected in the massive molecular clouds seen in these sources.

G. G. Fazio (ed.), Infrared and Submillimeter Astronomy, 127. All Rights Reserved.

INFRARED OBSERVATIONS OF EXTRAGALACTIC SOURCES

Douglas E. Kleinmann
Center for Astrophysics
Harvard College and Smithsonian Astrophysical Observatory
Cambridge, Massachusetts 02138

1. INTRODUCTION

The available balloon-borne and airborne infrared data on extragalactic sources, in particular M 82, NGC 1068 and NGC 253, will be reviewed and discussed in the context of the extensive groundbased work. The data will be examined for the clues they provide on the nature of the ultimate source of the energy radiated and on the mechanism(s) by which it is radiated.

Since the discovery of unexpectedly powerful infrared radiation from extragalactic objects -- a discovery now about 10 years old (see Low and Johnson 1965, Pacholczyk and Wisniewski 1967, Kleinmann and Low 1970a,b, and Rieke and Low 1972, *inter alia*) -- the outstanding problems in this field have been to determine
(1) the mechanism by which prodigious amounts of energy are released in the infrared, and
(2) the nature of the underlying energy source.

Data from air-, balloon-, and rocket-borne telescopes have helped to define the spectral distribution, luminosity, and energy balance of these extragalactic sources, thereby providing information fundamental to solving the two problems listed above. This "space experiment" data consist of broad bandpass photometry, most of it in the far-infrared (FIR). There is also limited information on source size.

The flow diagram given below outlines the procedure generally followed in analyzing this kind of data. The wide spectral bandpasses of the FIR filters (used in order to obtain adequate signal-to-noise ratios in suitable integration times) complicates the data analysis by making the effective wavelengths of the radiometry dependent on the intrinsic source distribution. Thus in analyzing the data a spectral distribution is initially assumed for the source,

G. G. Fazio (ed.), Infrared and Submillimeter Astronomy, 129-141.

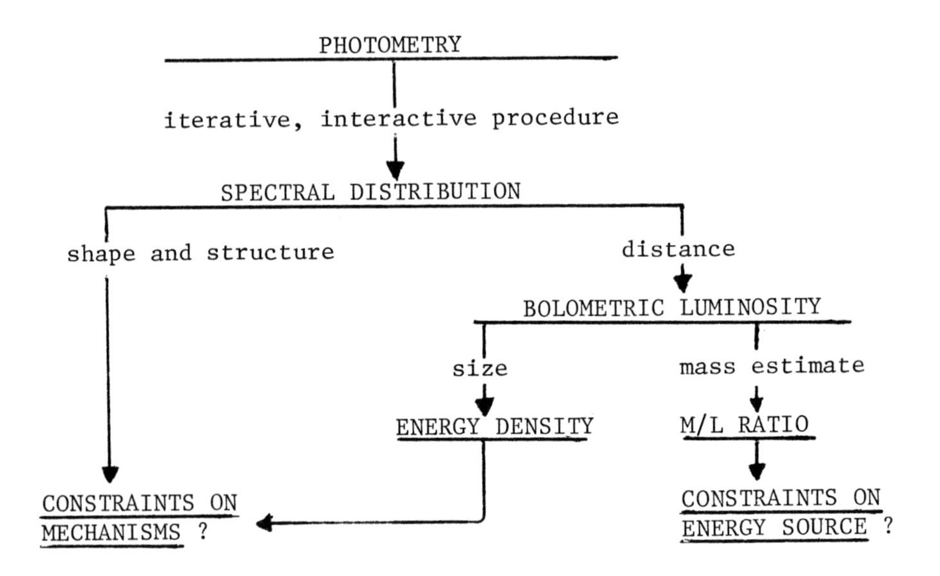

effective wavelengths calculated and fluxes derived, and a new spectral distribution defined; the process generally converges after one or two iterations. The final spectral distribution is examined for whatever constraints its overall shape and structure may impose on the radiative mechanism(s). The final spectral distribution is also integrated to obtain the total flux, and, if the distance to the source is known, the total (bolometric) luminosity. If, in addition, the size of the radiating volume is known and a mass estimate for that volume is available, the mass-to-luminosity ratio so obtained for the source may be useful in putting constraints on the energy source. The luminosity and size of the radiating volume can also be used to obtain the radiation energy density, which can be compared with the magnetic field energy density and mass to infer additional constraints on the radiation mechanism(s).

At this time only three extragalactic sources have been measured in the infrared using air-, balloon-, or rocket-borne telescopes. They are NGC 1068, M 82, and NGC 253. These three are not the only galaxies observed, and upper limits have been established for 7 others.

2. NGC 1068

NGC 1068 is the prototype for Class II Seyfert galaxies, in which the forbidden lines and the Balmer lines have the same widths

(Khachikian and Weedman 1974). Figure 1 gives the spectral distribution of NGC 1068. This figure follows that given by Telesco,

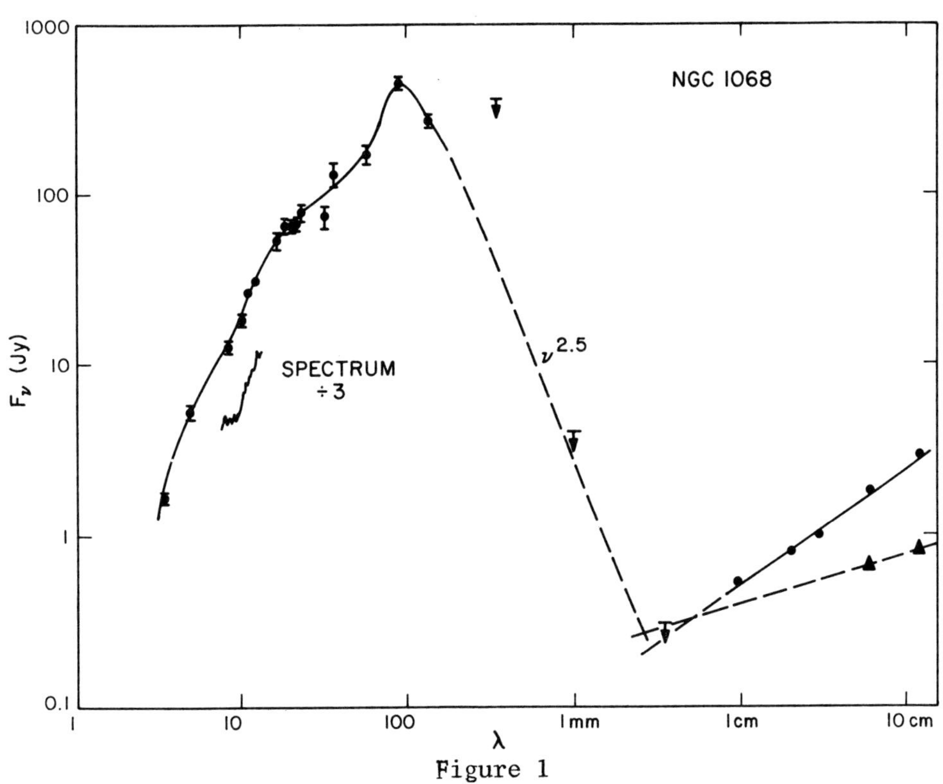

Figure 1

Harper, and Loewenstein (1976a). The 2 - 34μ photometry shown is that of Rieke (1975b), and the FIR data are those of Telesco et al. (1976a). The 8 - 13μ spectrum of Kleinmann, Gillett and Wright (1976) has been divided by 3 to avoid confusion with the bandpass photometry. The upper limit at 1 mm is from Elias et al. (1975), and the radio data are from Kellermann and Pauliny-Toth (1971). Also shown are radio observations from de Bruyn and Willis (1974) of a compact ($\leq$ 0.03") source in the nucleus. In this figure, and in Figures 3 and 6, the spectral distribution between the FIR peak and the high frequency radio data has been arbitrarily taken to be $\propto \nu^{2.5}$.

Integrating under the spectral distribution of Figure 1, and using a distance of 20 Mpc, one obtains a bolometric luminosity of 3.7×10^{11} $L_\odot$ for NGC 1068. The 10μ ($\Delta\lambda$ = 5μ) luminosity is 5×10^{10} $L_\odot$, or 14% of the total.

The spectral distribution of NGC 1068 is notable for the breadth and structure of its FIR peak. Telesco, Harper and Lowenstein (1976a) comment on the breadth of the spectral distribution of NGC 1068, noting that although the objects observed in our Galaxy require diverse temperatures, nevertheless that no one of them requires so wide a range in temperatures as would be required to synthesize the spectrum of NGC 1068. They also note that the spectral distribution of NGC 1068 appears to be wider than those of M 82 or NGC 253. Although the spectral resolution is low, it is quite evident that the spectral distribution of NGC 1068 has broad inflections and structure; somewhat higher spectral resolution may reveal details such as those evident in the 8 - 13μ spectrum ($\lambda/\Delta\lambda \simeq 50$) of Kleinmann, Gillett, and Wright (1976). This spectrum, which is reproduced in Figure 2, shows a broad "silicate" absorption at ~ 10μ, and the 12.8μ line of NeII in emission; as will

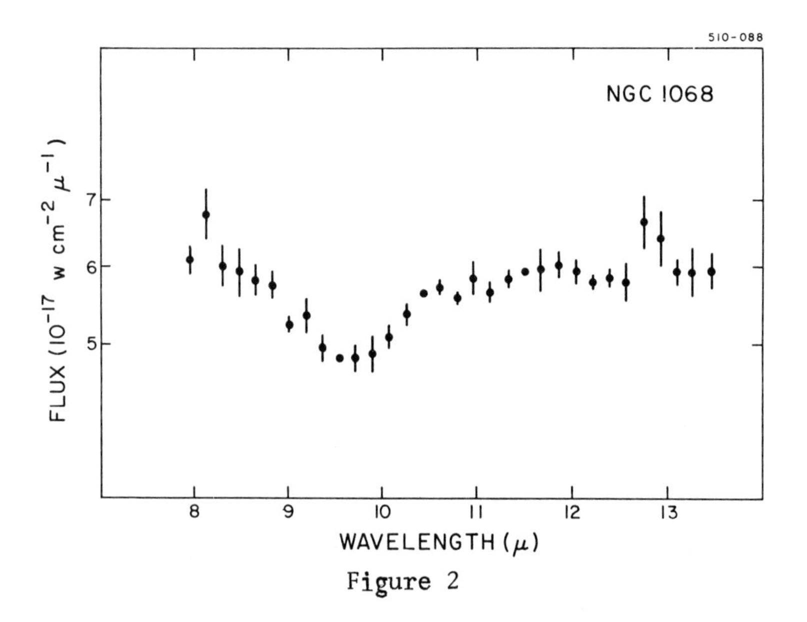

Figure 2

be seen, these features are also common to the other two galaxies now known to be powerful radiators in the FIR.

3. M 82

The spectral distribution for M 82 is shown in Figure 3. The 2.2, 3.5, 11.6, and 19μ observations were made with 60-75" beams on the KPNO 1.3-m and Mt. Hopkins Observatory 1.5-m reflectors. The letter "A" denotes the 11μ and 20μ fluxes for the 3' × 10' beam of the AFCRL rocket-borne infrared sky survey (Walker and Price 1975); note that M 82 is the only identified extragalactic source in the catalog of that survey. The 8-13μ spectrum was obtained by Gillett, Kleinmann, Wright and Capps (1975) using a 7" beam, and the "silicate" absorption and NeII line emission are clearly evident. The 69μ data was obtained by Kleinmann, Wright, & Fazio (1969) with the

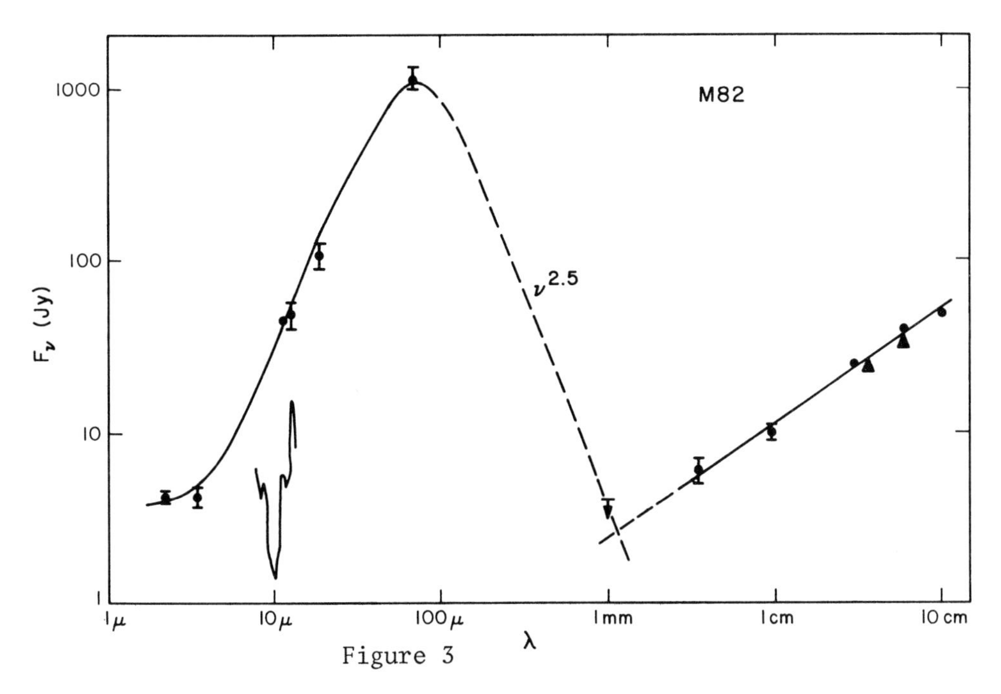

Figure 3

CfA/U of A 102 cm balloon-borne telescope. Additional FIR data, consistent with the 69μ datum shown, have been obtained by several of the various groups that have flown on NASA's Kuiper Airborne Observatory, including Harvey and Hoffmann of the University of Arizona, Telesco, Harper and Lowenstein of Yerkes, and Werner, Backlin, and Gatley of Galtech. The radio spectrum is from Kellermann and Pauliny-Toth (1971), and the 1 mm upper limit is from Elias et al. (1975). The filled triangles represent the fluxes from the areas mapped in the high spatial resolution radio maps at 5 GHz (Hargrave 1974) and at 8 GHz (Kronberg and Wilkinson 1975).

M 82 has been resolved at 10μ (Kleinmann and Low 1970b) and the effect of its size is evident in the mid-infrared data of Figure 3. Maps with angular resolutions of 2-5" and spectral resolving power (λ/Δλ) between 2 and 10 have been produced by Rieke and Low (in preparation) at several mid-infrared wavelengths; using Gillett's filter wheel spectrometer with a spectral resolving power of ~ 50, Kleinmann and Wright (in preparation) have mapped the spatial distribution of the line radiation from Ne^+ at 12.8μ and from Bracket α at 4.05μ. Because of the small beams and differential chopping employed to make these maps, regions of low surface brightness, or of small gradients in surface brightness are discriminated against in these maps, and the source size may be suppressed. The data from a simple beam-size experiment are shown on the right-hand side of Figure 4; the range of acceptable fits is shown on the left. The data shown include a series of measurements made on the Mt. Hopkins Observatory 1.5-m reflector using a bolometer system

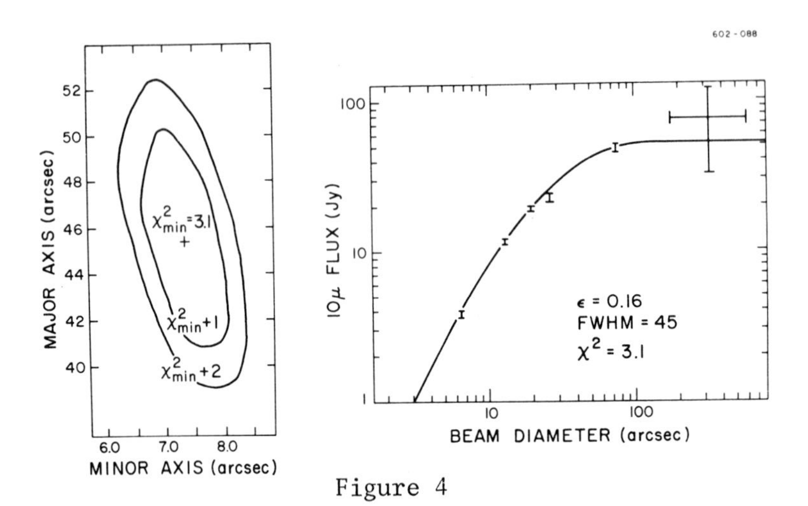

Figure 4

featuring cooled interchangeable focal plane apertures, a 75" beam measurement made on the KPNO 1.3-m telescope with a high throughput photometer, and the 11μ observation from the AFCRL Catalogue (Walker and Price 1975). The error bars shown are 1σ and reflect only the statistics of the measurement; since no error bars were given for the AFCRL data, a $0^m.5$ error was assigned. The range of possible fits shown on the left of Figure 4 assumes that the surface brightness follows a gaussian distribution along each axis. The contour $\chi^2_{min} + 1$ indicates 1σ changes in the parameters of the fit, and circumscribes the range of acceptable fits. With 2 degrees of freedom, the best fit has $\chi^2 = 3.1$, indicating that the assumption of a gaussian distribution in surface brightness provides a reasonable fit. Thus at 10μ the FWHM along the major and minor axes of the M 82 are 45" and 7.5" respectively, with 1σ uncertainties of 10% in each dimension. The performance of these beam size measurements also showed that the 10μ centroid was ~ 10" SW of the position given by Kleinmann and Low (1970b).

A FIR beam size effect has been noted for M 82 by Werner _et al._ (1976) and by Telesco _et al._ (1976b). Each obtains somewhat more flux in a 1' beam than in a 30" beam. These results support the conclusion drawn from the scan data made with a 1' × 1.5' beam by Kleinmann, Wright, and Fazio (1976) at an inclination of 26° to the plane of M 82. The data for M 82 are shown as filled circles in Figure 5; the solid lines show comparison scans of Mars, which was 70× brighter. By fitting gaussian curves to the observations of Mars and of M 82, allowing for the shift in effective wavelength between the two, and assuming the source has the same ellipticity in the FIR that is observed in the optical, mid-infrared, and radio, it was concluded that the size of the FIR source along the plane of M 82 has a 90% probability of being in the range 46-76", and that

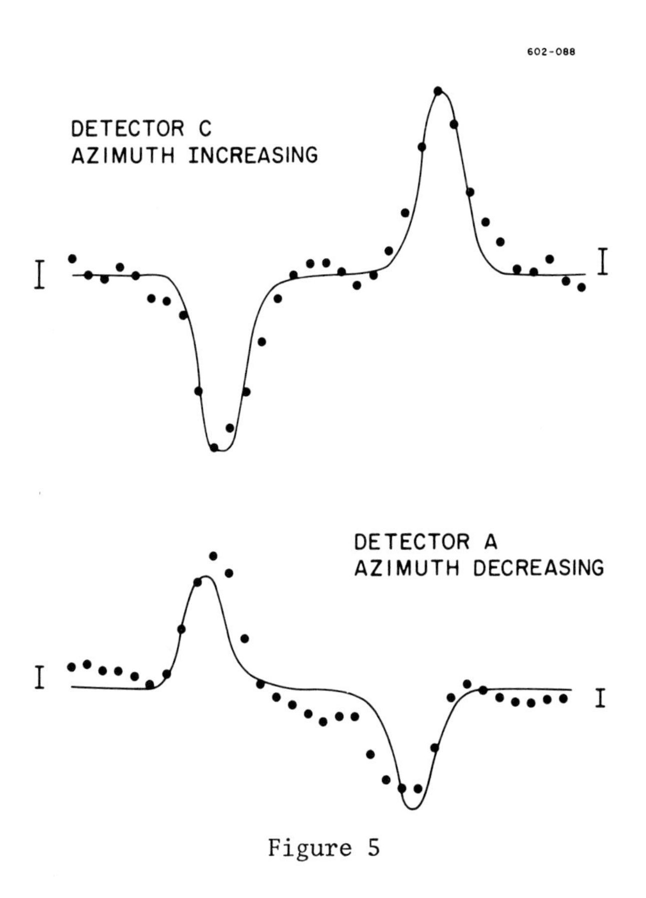

Figure 5

the most likely value is 63". On the other hand, Telesco _et al._ conclude from scans made through the peak of the source at an inclination of 15° to the plane of M 82, that at 60μ the source measured in M 82 is probably 40-45" (or 41-47" along the plane). It is reasonable to conclude that the FIR emitting volume in M 82 is at least as large as the 10μ emitting volume, and that it may well be larger.

The bolometric luminosity of M 82 is 3×10^{10} $L_\odot$ if the galaxy is at 3.2 Mpc. The 10μ luminosity determined from the 75" beam observation is 2×10^9 $L_\odot$, or 7% of the total luminosity.

The mass estimated for the central 63" ($\geq$ 45") from the emission line rotation curve of Burbidge, Burbidge and Rubin (1964) is 13 ($\geq$ 8) $\times 10^8$ $M_\odot$. Thus, the mass-to-luminosity ratio for M 82 in solar units, is 0.04 ($\geq$ 0.03). This result is significantly different from that of Harper and Low (1973), although the two can be reconciled: the flux values used by Harper and Low are a factor of 2 to 2.5 too large, the volume they assumed is between 3 and 9 times too small, and the use of the absorption line rotation curve underestimates the mass by an additional factor of 4.8.

4. NGC 253

NGC 253 is the only other galaxy measured with one of the "space experiments" being discussed at this meeting. The spectral distribution of this edge-on Sc galaxy is shown in Figure 6. The 2-20μ data are drawn from Becklin, Fomalont and Neugebauer (1973)

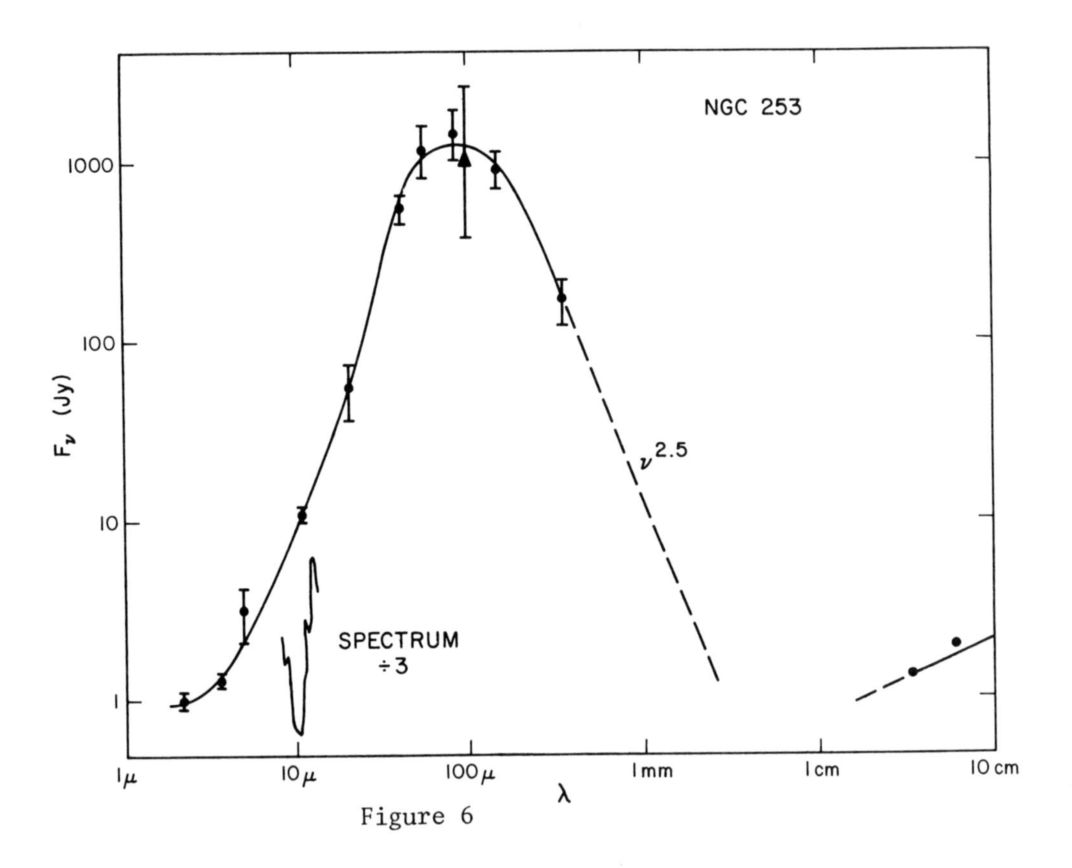

Figure 6

and Rieke and Low (1975a). NGC 253, like M 82, is extended at mid-infrared wavelengths, but the data suggest that the source is smaller than 20". The 2-20μ data shown in Figure 6 are those appropriate for 20-30" beams. The 8-13μ spectrum ($\lambda/\Delta\lambda \sim 50$) of Gillett, Kleinmann Wright, and Capps (1975) has been divided by 3 to avoid confusion. Like those of NGC 1068 and M 82, this spectrum shows the silicate absorption and NeII in emission; moreover, the 8-13μ spectra of NGC 253 and M 82 are strikingly similar both in what they contain and in the contrast with which they show it. Except for the 100μ datum, the 40-150μ observations have been generously provided in advance of their publication by Telesco et al. (1976b), who have also found a $\leq 1\sigma$ increase between the 58μ fluxes measured with 30" and 60" beams. The 100μ observation was published by Harper and Low (1973). The 350μ datum is found in Rieke et al. (1973). The 3.5 cm and 6 cm radio data were obtained from Shimmons and Wall (1973) and Whiteoak (1970). Additional longer wavelength data are given by Becklin et al. (1973), Fomalont (1971), and Slee (1972).

The bolometric luminosity of NGC 253 is 2×10^{10} $L_\odot$. The 10μ luminosity is 4×10^8 $L_\odot$, or 2% of the total Assuming that the far infrared flux comes from the same central 20" diameter volume as the mid-infrared flux, Rieke and Low (1975a) conclude that the mass-to-luminosity ratio is ≈ 0.03.

Although the M/L ratios are quite low for NGC 253 and M 82, they do not require that the energy source be non-stellar. Figure 7 shows the fraction of the mass that must be converted into stars in order to obtain a particular M/L ratio. The distribution of these stars is assumed to follow the Salpeter (1959) "initial" luminosity

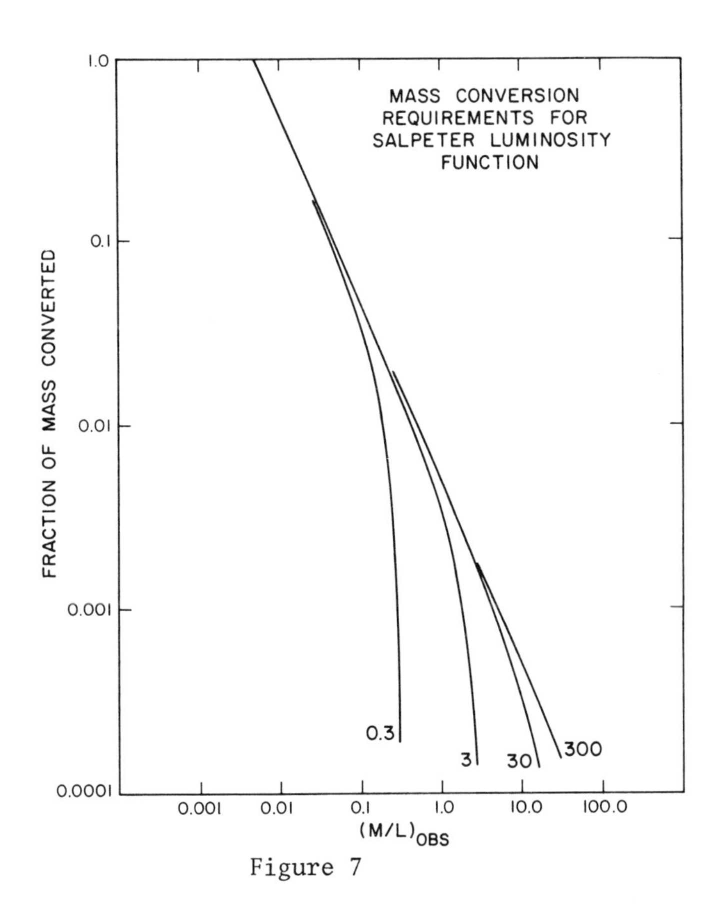

Figure 7

function. The curves are labelled by the mass-to-luminosity ratio that obtained when the conversion began. This figure shows that the minimum obtainable M/L for the Salpeter luminosity function is 0.005; lower values would require either extrapolation beyond $M_V = -5$ stars, or that the luminosity function be skewed toward higher mass, higher luminosity stars. Such a skewing might occur for example in a relatively warm interstellar gas, which would require larger masses before collapse could commence. Figure 7 also shows that for $M/L \leq 0.1$ the mass fraction that must be converted is insensitive to the initial M/L; in particular, for $M/L \leq 0.1$,

$$(\text{mass fraction}) \times (M/L) \simeq 0.005.$$

Thus, the M/L ratios for M 82 and NGC 253 are not too low for the energy source to be thermonuclear reactions, provided that 10-15% of the mass is being or has recently been converted to stars. However, as Rieke and Low (1975a) pointed out for NGC 253, a $M/L \simeq 0.03$ is too low to be supported by thermonuclear reactions continuously for a normal galactic lifetime. Thus, the presently observed radiation is either transient (and perhaps recurrent), or mass is replenished at the nucleus, or the energy is produced in another, more efficient process.

The M/L ratio for NGC 1068 is certainly smaller than those for M 82 and NGC 253, and it may be small enough to require energy from a source other than thermonuclear burning. Mass estimates for angular radii $\leq 40''$ are only poorly determined because of the mass motion in the nuclear region (Walker 1968). Using the 1' beam observations of Telesco et al. (1976a) and Walker's mass estimate for the central 80" diameter volume, $M/L \simeq 0.016$. Although the FIR size is not necessarily as small as the 1" size determined at 10μ by Becklin et al. (1973), it is apparently much smaller than 80": the 1' beam 38μ measurement (132±22 Jy) of Telesco et al. (1976a) and the 4" beam 34μ measurement (72±19 Jy) of Rieke and Low (1975b) differ by less than a factor of 2, although that difference is 2.5σ. Arguing that the mass distribution follows that of surface brightness, Rieke and Low (1975b) conclude that $M/L \leq 0.003$ for NGC 1068, and this conclusion is reinforced by Telesco et al. (1976a). By comparison, the smallest value of M/L that can be supported by stars following a Salpeter luminosity function, even after that function has been extended extrapolated to $M_V = -7$, is $M/L = 0.003$.

5. SUMMARY OF MEASURED OBJECTS

None of the extragalactic objects measured in the FIR is "normal". All have substantial amounts of dust that is evident both in optical photographs and spectrograms, and in the infrared "silicate" absorption feature. They are the brightest of the galaxies measured at 10μ. Each has been resolved at 10μ and a size has been measured. The spectral distribution of each rises rapidly into, and peaks in the FIR, and the bolometric luminosities are dominated by the contributions in the FIR. None of these galaxies present compelling evidence for a non-thermal radiation mechanism, or for a non-thermonuclear energy source, although none can continue to radiate via thermonuclear reactions at their present levels for a normal galactic lifetime.

6. OTHER OBSERVATIONS

Upper limits have been obtained on 7 other extragalactic sources. Observations by Telesco et al. (1976b) place 2σ upper limits of ~ 50 Jy at 100μ on the class I Seyfert prototype NGC 4151, NGC 4705, M 51, M 33, the archetype of normal galaxies M 31, and the most luminous galaxy MKN 231. A similar limit was obtained for MKN 231 by Harvey and Hoffman (1976). Telesco et al. (1976b) have also obtained a 2σ upper limit of 4 Jy at 100μ on 3C 273.

Since the radiation from the NGC 1068, M 82, and NGC 253 appears to be explicable in terms of thermal mechanisms, it is interesting to examine the spectral distributions of sources for which the mechanisms may be non-thermal. Such a prejudice is usually entertained for the three objects whose spectral distributions are shown in Figure 8: class I Seyferts (e.g. NGC 4151), quasi-stellar objects (e.g. 3C 273), and the luminous, rapidly varying

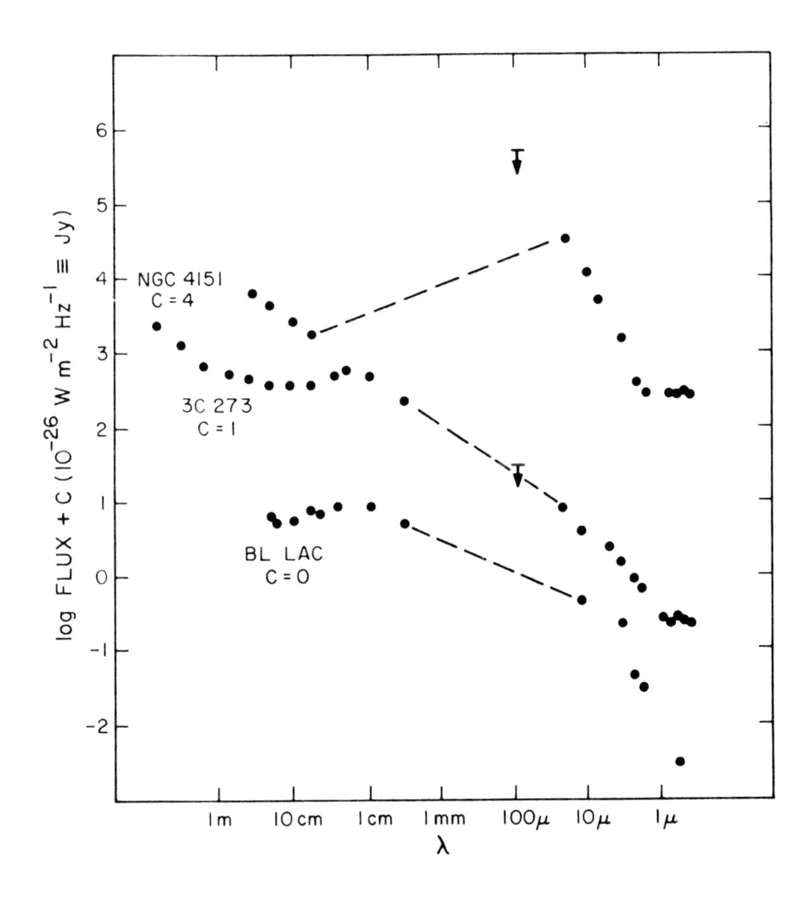

Figure 8

lacertids (e.g. BL Lac). Figure 8 shows that 3C 273 is very much different from NGC 1068, M 82, and NGC 253 in that its spectral distribution does not peak at ~ 60μ, and the bolometric luminosity is dominated by the radiation at $\lambda \leq 50\mu$. NGC 4151 may also peak at a wavelength as short as 35μ; the ratio between its bolometric and 10μ luminosities also appears to be somewhat lower than for NGC 1068.

The ratio of the total luminosity to the 10μ luminosity reflects both the shape and breadth of the spectral distributions, and the temperatures that characterize them. Rieke and Low (1975c) suggested that this ratio might have a "typical" value of ~ 15, within a factor of 2 or 3. Although only 3 galaxies have been measured in the FIR, the range in values they provide for this ratio is already enough to strain this suggestion of Rieke and Low, and the upper limit for 3C 273 by Telesco et al. (1976b) tends to invalidate it altogether. With additional observations it should be possible to test whether this ratio is systematically lower for "nonthermal" sources, or whether it decreases with increasing luminosity, both of which are weakly suggested by the present data.

Future use of the Infrared Astronomy Satellite (IRAS), the Shuttle Infrared Telescope Facility (SIRTF), the Large Space Telescope (LST), and the Large Infrared Telescope for the Shuttle (LIRTS) --combined with the airborne and balloon-borne systems now being developed-- will provide a panoply of exciting observations and observational techniques for extragalactic infrared astronomy. Space-borne facilities will provide a combination of potentially long integration times and an absence of the atmosphere that will result in vastly improved photometric precision throughout the infrared. Infrared uses for these facilities will include application of increased spatial and spectral resolution, determination of temporal behavior, measurement of polarization, and statistical studies of infrared emitting galaxies.

REFERENCES

Becklin, E.E., Fomalont, E.B. and Neugebauer, G. 1973, Ap. J. (Letters), 181, L27.

Becklin, E.E., Matthews, K., Neugebauer, G. and Wynn-Williams, G.C. 1973, Ap. J. (Letters), 186, L69.

de Bruyn, A.G. and Willis, A.G. 1974, Astron. and Astrophys., 33, 351.

Burbidge, E.M., Burbidge, G.R. and Rubin, V.C. 1964, Ap. J., 140, 942.

Elias, J.H., Gezari, D.Y., Hauser, M.G., Neugebauer, G., Werner, M.W., Westbrook, W.E. 1975, Bull. Am. Ast. Soc., 7, 436.

Fomalont, E.B. 1971, Ap. J., 76, 513.

Gillett, F.C., Kleinmann, D.E., Wright, E.L., and Capps, R.W. 1975, Ap. J. (Letters), 198, L65.

Hargrave, P. J. 1974, M.N.R.A.S., 168, 4 1.

Harper, D.A. and Low, F.J. 1973, Ap. J. (Letters), 182, L89.

Harvey, P.M. and Hoffmann, W.F. 1976, private communication

Kellermann, K.I. and Pauliny-Toth; IIK 1971, Astrophysical Letters, 8, 153.

Khachikian, E. Ye, and Weedman, D.W. 1974, Ap. J., 192, 581.

Kleinmann, D.E. and Low, F.J. 1970a, Ap. J. (Letters), 159, L165.

Kleinmann, D.E. and Low, F.J. 1970b, Ap. J. (Letters), 161, L203.

Kleinman, D.E., Gillett, F.C. and Wright, E.L. 1976, Ap. J., in press.

Kleinmann, D.E., Wright, E.L. and Fazio, G.G. 1976, CfA preprint No. 500.

Kronberg, P.O. and Wilkinson, P.N. 1975, Ap. J., 200, 430.

Low, F.J. and Johnson, H.L. 1965, Ap. J., 141, 336.

Pacholczyk, A.G. and Wisniewski, W.Z. 1967, Ap. J., 147, 394.

Rieke, G.H. and Low, F.J. 1972, Ap. J. (Letters), 176, L95.

Rieke, G.H. and Low, F.J. 1975a, Ap. J., 197, 17.

Rieke, G.H. and Low, F.J. 1975b, Ap. J. (Letters), 199, L13.

Rieke, G.H. and Low, F.J. 1975c, Ap. J. (Letters), 200, L67.

Rieke, G.H., Harper, D.A., Low, F.J. and Armstrong, K.R. 1973, Ap. J. (Letters), 183, L67.

Salpeter, E.E. 1959, Ap. J., 129, 608.

Shimmons, A.J. and Wall, J.V. 1973, Australian J. Phys., 26, 93.

Slee, O.B. 1972, Astrophysical Letters, 12, 75.

Telesco, C.M., Harper, D.A., and Loewenstein, R.F. 1976a, Ap. J. (Letters), 203, L53.

Telesco, C.M., Harper, D.A. and Loewenstein, R.F. 1976b, private communication.

Walker, M.F. 1968, Ap. J., 151, 71.

Walker, R.G. and Price, S.D. 1975, AFCRL TR 75 0373.

Werner, M.W., Becklin, E.E. and Gatley, I. 1976, private communication.

Whiteoak, J.B. 1970, Astrophysical Letters, 5, 29.

SPECTRUM AND ISOTROPY OF THE SUBMILLIMETER BACKGROUND RADIATION

Dirk Muehlner

Physics Department and Research Laboratory
of Electronics
Massachusetts Institute of Technology
Cambridge, Massachusetts 02139

1. INTRODUCTION

Two great astronomical discoveries have most shaped our present concept of the Big Bang universe. Like the Hubble recession of the galaxies, the discovery of the 3°K background radiation by Penzias and Wilson [1] in 1965 has given rise to a line of research which is still very active today. Penzias and Wilson's universal microwave background at 7 cm was immediately interpreted by R. H. Dicke's group [2] at Princeton as coming from the primordial fireball of incandescent plasma which filled the universe for the million years or so after its explosive birth. This interpretation gives rise to two crucial predictions as to the nature of the background radiation. Its spectrum should be thermal even after having been red shifted by a factor of ~1000 by the expansion of the universe, and the radiation should be isotropic--assuming that the universe itself is isotropic. If the background radiation is indeed from the primordial fireball, it affords us our only direct view at the very young universe. This paper will deal with the spectrum and then the isotropy of the background radiation, with emphasis on high frequency or submillimeter measurements. Prospects for the future will be discussed briefly at the end.

2. SPECTRUM

By about 1968 the microwave region of the background spectrum had been fairly well covered by ground-based measurements. The results were consistent with the Rayleigh-Jeans tail of a blackbody spectrum with a temperature of roughly 2.7°. There was even some evidence that the spectrum was beginning to show the proper

G. G. Fazio (ed.), Infrared and Submillimeter Astronomy, 143-152.

deviations from ν^2 at frequencies approaching the blackbody peak at $\lambda^{-1} \simeq 6$ cm^{-1}. The best early evidence that the background spectrum was thermal up to 3.8 cm^{-1} was provided by interstellar cyanogen acting as a remote thermometer for the background temperature. The fascinating work on interstellar molecules is beyond the scope of this paper; however, see Refs. [3,4].

In the several years after the Rayleigh-Jeans tail of the background spectrum had been satisfactorily measured, the roughly thermal nature in the submillimeter region (here taken as starting at roughly 2 cm^{-1}) has been established by much painstaking effort. In particular, the spectrum has been shown to turn over and decrease beyond the black body peak. Despite the great amount of effort which has been put into the submillimeter measurements,

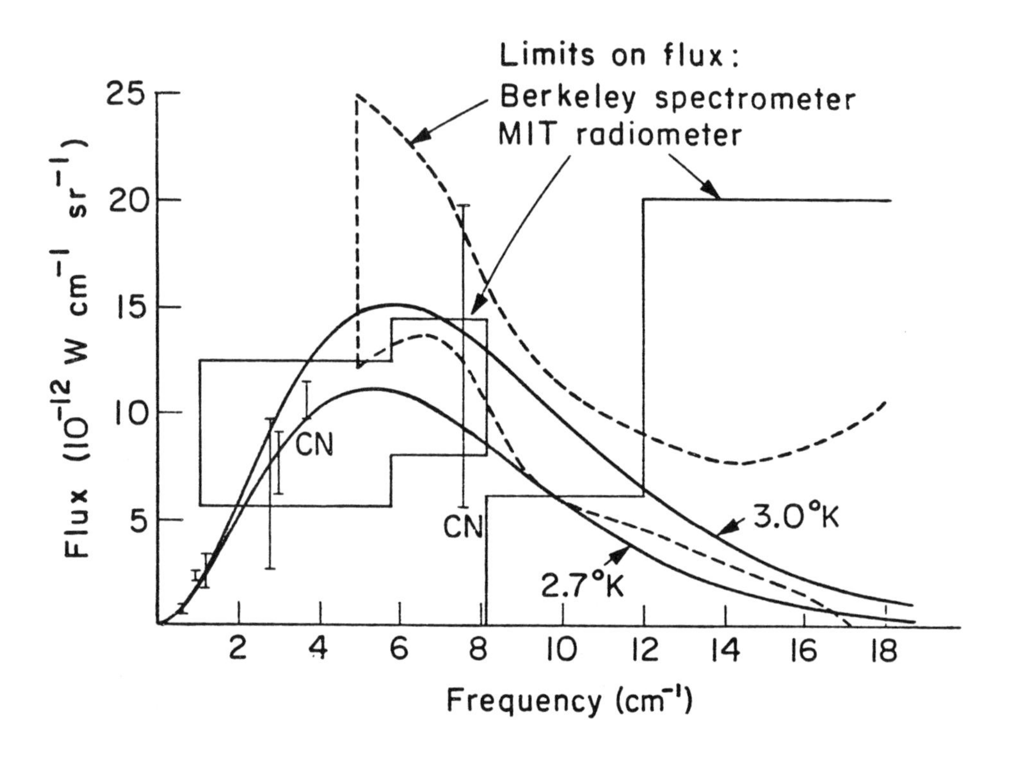

Fig. 1. Limits on the submillimeter cosmic background spectrum. CN refers to interstellar cyanogen. A few high frequency microwave points are also shown.

this work is still far from completed. The importance of the submillimeter spectrum is obvious: the low-frequency power low spectrum might be explained in different ways, but a complete blackbody spectrum is the clear signature of the primordial fireball. Furthermore, slight deviations from a perfect thermal spectrum (or perfect isotropy, as will be discussed later) would give invaluable information about the early evolution of the universe.

Why has the submillimeter region proved so recalcitrant to precise spectral measurements? There are two overriding reasons. The broadband far-infrared detectors that have been used for this spectral region are in a primitive state of development compared to the highly sensitive radio receivers used at lower frequencies. Even more important, the atmosphere becomes increasingly opaque at wavelengths shorter than one centimeter, and the powerful foreground noise it generates frustrates attempts to measure the weak background even from mountaintops or aircraft. With the exception of one measurement made in an atmospheric "window" at about 9 cm^{-1} [5], all of the submillimeter measurements of the background spectrum have been made by infrared radiometers or spectrometers carried above the bulk of the atmosphere by balloons or rockets. An added complication at submillimeter frequencies is that beyond the 3°K black body peak, where the background spectrum decreases exponentially, it will be completely overwhelmed by thermal radiation from any warm radiometer parts, which still increases as ν^2. Thus it has been necessary to use instruments with all components in the optical train, as well as the detectors themselves, cooled to liquid helium temperature.

The difficulties that faced the first submillimeter background measurements are dramatized by the fact that the first measurement, made from a rocket flown by Shivanandan, Houck, and Harwit [6] of the Naval Research Laboratory and Cornell, indicated a flux about fifty times larger than was expected from a 2.7° black body, and the second, from a balloon-borne radiometer built by Muehlner and Weiss [7] of MIT, found an only slightly smaller flux in one spectral band out of three. That these early results were spurious was shown in 1971 in a rocket flight by the Los Alamos group of Blair *et al*. [8] and additional balloon flights by Muehlner and Weiss [9]. The results of the 1971 flights were consistent with a 2.7°K black body, and the MIT instrument, which had a filter wheel to provide spectral information, showed for the first time that the spectrum turned over above the black body peak at about 6 cm^{-1}.

Since the 1971 balloon flights the major advance in the direct measurement of the background spectrum has been made by Woody, Mather, Nishioka, and Richards of Berkeley [10]. They flew a liquid helium-cooled polarizing Fourier transform spectrometer using a Ge bolometer, from a balloon at an altitude of 39 km, and have established that the spectrum of the background approximates that of a 2.99°K black body between ~6 and ~16 cm^{-1}. Data from a more recent flight of this spectrometer are still being analyzed. The group of Robeson et al. of the University of London have also flown a balloon-borne Fourier spectrometer and obtained data consistent with a 2.7°K black body [11]. P. Marsden's group at Leeds University, U.K. have built a balloon-borne lamellar grating interferometer, but have not yet obtained useful data.

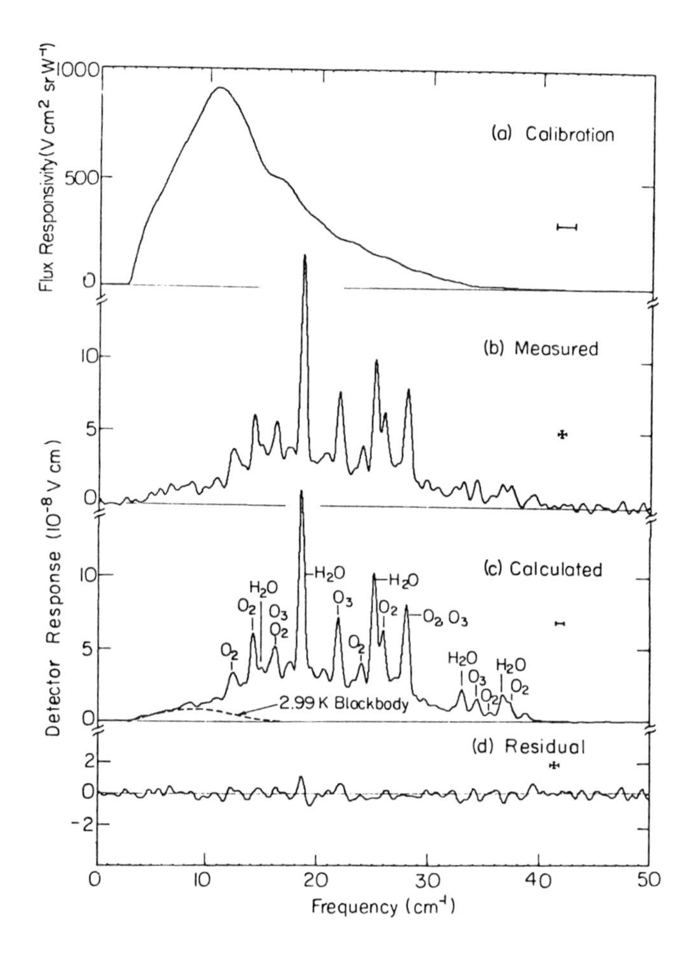

Fig. 2. Results from the Berkeley spectrometer showing atmospheric emission at ~40 km altitude (from Ref. 10). (a) Instrumental flux responsivity as a function of frequency. (b) Observed instrumental response to the night sky. (c) The fitted-model spectrum. The origins of some of the stronger atmospheric emission lines are shown. (d) The difference between the curves of (b) and (c).

Figure 1 summarizes the current state of the submillimeter measurements of the background spectrum. Several measurements which constrain the spectrum less well than those shown have been omitted. A glance at this figure will convince the reader that while the background spectrum has been shown to be roughly thermal with a temperature of 2.7-3°K, its exact shape is far from well determined. The problem which ultimately limits the accuracy of measurements of the background radiation even at balloon altitudes is the residual atmospheric emission, which makes it impossible to achieve the limits set by detector noise. The relatively large uncertainties in the measured spectral flux densities shown in Fig. 1 are due to uncertainties in subtracting the atmospheric contribution to the total measured flux. Figure 2, from the paper of Woody *et al.*, shows dramatically the extent to which atmospheric emission overpowers the 3°K black body even at 40 km altitude, especially at the higher frequencies which are of special interest.

3. ISOTROPY

The isotropy--or more properly the degree of anisotropy--of the cosmic background radiation is of as much interest as its spectrum in the information it must carry about the oldest and largest sample of the universe available to us. Two types of anisotropy are usually considered. Granularity, or small-scale variations over angles of the order of minutes of arc or less, could result if the background radiation were due to a superposition of many discrete sources (e.g., galaxies) in the sky, a possibility which measurements have essentially ruled out. A more likely source of granularity is condensation in the primeval plasma which later gave rise to galaxies or clusters of galaxies. Very stringent limits have been set on the small-scale anisotropy at microwave wavelengths [12], but as yet no serious plans are under way to measure the submillimeter granularity of the background. Large-scale anisotropies, generally taken to refer to scales of $\sim 10°$ or larger, may also have several causes. The universe may not have exploded isotropically, or the radiation may have been perturbed by interaction with mass density inhomogeneities of supercluster size. In any event, the isotropy of the background will provide our best evidence for or against the Cosmological Principle that the universe in the large is homogeneous and isotropic. One source of large-scale anisotropy *must* exist--a dipole or 24-hour term due to the motion of the Earth with respect to the primordial fireball, i.e., the "absolute motion" of the Earth. The peculiar velocity of the Earth, which is the vector sum of the orbital motion of the Earth around the sun, the sun around our Galaxy, and the motion of our Galaxy itself, is of course of considerable interest in astronomy, and perhaps to future generalized Michelson-Morley experiments aimed at detecting the effect on local physics of the existence of a universal rest frame. The velocity anisotropy has the form $\Delta T \simeq T_0(1+(v/c)\cos\theta)$ where θ is the angle between the observation direction and the velocity of the Earth. The velocity v is expected to be of the order of 10^{-3} c, so the expected anisotropy amounts to temperature differences of millidegrees between different parts of the sky. Since this effect necessarily exists if the whole picture of the background radiation is correct, it sets a natural goal for the sensitivity of any large-scale isotropy experiment.

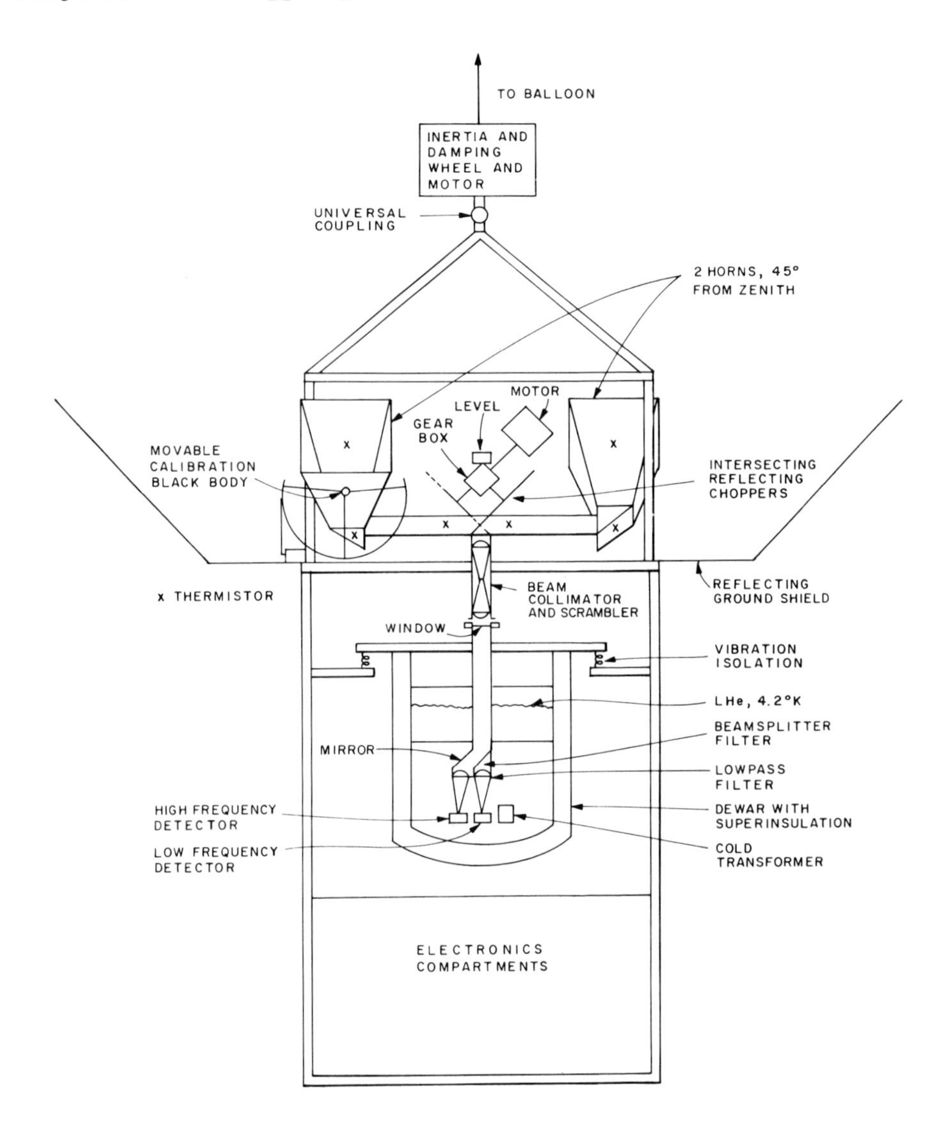

Fig. 3. The MIT balloon-borne submillimeter differential radiometer for measuring the background anisotropy.

Several large-scale anisotropy measurements have been made at microwave frequencies [13]. These experiments require difficult corrections to the observed data. At long wavelengths the galactic synchrotron background dwarfs any expected anisotropy in the cosmic background, and at higher frequencies where the galactic background is small enough to be subtracted with some confidence, atmospheric emission is a severe problem. More recent microwave measurements have therefore been made from balloons. The latest microwave measurement is that of Corey and Wilkinson [14] at 19 GHz. Their preliminary results indicate an Earth velocity of order 10^{-3} c, different from zero by one or two standard deviations. The measurement of the microwave anisotropy is presently an active field, and several experiments are in preparation, including a 33 GHz experiment to be flown on the NASA U-2 aircraft in the near future [15] and one planned for a medium duration trans-Atlantic balloon flight [16]. This effort is of special interest as a first step toward using the developing technology of globe-circling super-pressure balloons, which will provide weeks of observing time.

If the large-scale anisotropy due to the Earth velocity can be measured by microwave receivers, why bother with difficult submillimeter measurements? One reason is that for anisotropies other than that due to Earth motion, which merely modulates the temperature of a thermal spectrum in different directions, the spectra of any intrinsic background anisotropies will themselves be of great interest. More important, in view of contribution to the anisotropy of "local" sources such as galactic synchrotron emission at low frequencies and thermal radiation by galactic dust at high frequencies, it will be essential to obtain measurements in as many bands as possible. This will help to distinguish that portion of the observed anisotropy actually due to the primordial fireball from contributions of nearer (still interesting!) sources.

The only submillimeter anisotropy experiment at present is that of Muehlner and Weiss at MIT. The experiment employs a balloon-borne dual channel differential radiometer (Fig. 3) with bandwidths of 2-10 and 10-30 cm^{-1}. This instrument has been described in Ref. [17]. The low frequency channel is sensitive primarily to the 3°K background, and the high frequency channel monitors the atmosphere so that corrections may be made for residual atmospheric contributions to the low frequency signal. The fact that the high frequency channel is sensitive to radiation from dust in the galactic disk presents a problem. To make the required (small) correction for this galactic background, it may be necessary to have a low resolution map of the galactic submillimeter brightness and spectrum. The MIT group has begun making such a map with a version of the isotropy instrument modified for a 1.6° beamwidth [18].

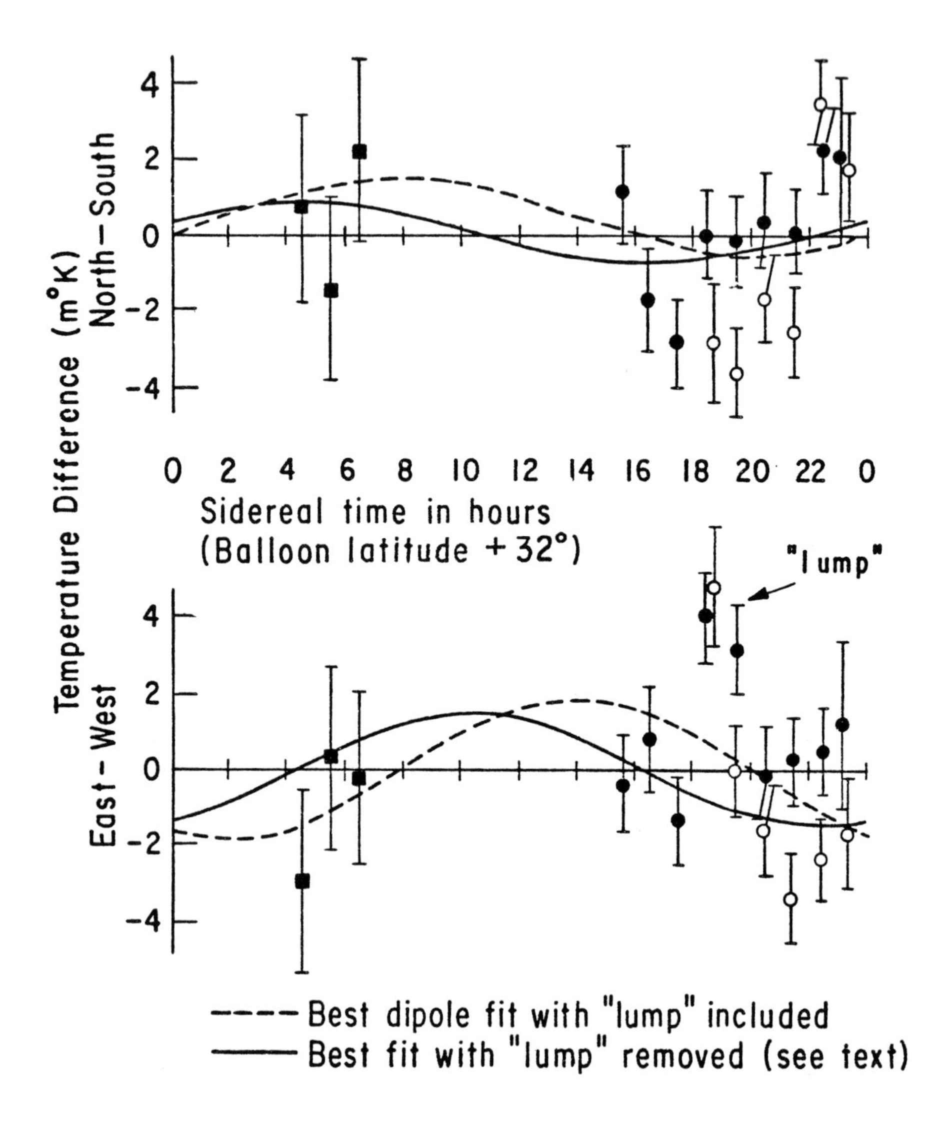

Fig. 4. MIT submillimeter large-scale isotropy data from three balloon flights: January 1974; ■ June 1974 ● ; August 1974 ○ .

Four flights of this differential radiometer have been made, of which the first three, in January, June, and August of 1974 were successful and the last, in July 1975, produced no useful data. The sensitivity of the experiment to a difference in temperature between the two beams with the InSb detectors used so far has been $\sim 0.03°$ $kHz^{-1/2}$ (antenna temperature). This sensitivity should allow the velocity of the Earth to be determined with a (1σ) error of ~ 75 km/sec in seven hours of observing time, which was available in the June flight alone.

Figure 4 shows the North-South and East-West components of the low frequency signal from the 1974 flights. The data have been averaged into hourly bins, and are corrected for atmospheric effects (amounting typically to a fraction of a millidegree). Probably as a result of systematic errors in the data which are not yet fully understood, no assumed Earth velocity gives a fit with a reasonable confidence level. However, manipulations of the data such as removing the obvious "lump" in the East-West signal or omitting the atmospheric corrections all are consistent with a velocity of 200 km/sec in the direction RA $\simeq 18^h$, DEC ~ 0, with an uncertainty (90% confidence) of 300 km/sec in any direction.

The most important result of both microwave and submillimeter large-scale anisotropy measurements to this time is that the temperature of the cosmic background varies by no more than about one part per thousand over large angles. This already provides our best evidence that the universe in the large is isotropic.

The MIT group is planning more flights in the imminent future with an improved apparatus. The most important change from previous flights is that the InSb detectors are being replaced by composite bolometers which are $\sim 10^2$ times more sensitive.

4. FUTURE

Measurements of the spectrum may still benefit from balloon flights to higher altitudes than have been achieved so far, and both submillimeter spectrum and isotropy will benefit from the long observing times which will become available with long duration super-pressure balloons. High precision measurements of the spectrum and of the anisotropy at a variety of frequencies await the availability of the combination of very long observing times, freedom from the atmosphere, and complete sky coverage which only satellites can offer. A proposal to mount a small submillimeter spectrum experiment on the secondary mirror of the IRAS telescope has been considered, but was judged to interfere with the primary survey mission of IRAS.

However, NASA has initiated a study of a dedicated Cosmic Background Explorer Satellite which will include both spectrum and isotropy experiments at several wavelengths.

References

1. A. A. Penzias and R. W. Wilson, Astrophys. J. 142, 419 (1965).

2. R. H. Dicke, et al., Astrophys. J. 142, 414 (1965).

3. P. Thaddeus, Annu. Rev. Astron. Astrophys. 10, 305 (1972).

4. D. J. Hegyi, W. A. Traub, and N. B. Carleton, Astrophys. J. 190, 543 (1974).

5. Dall'Oglio, et al., Paper at Millimeter Wave Conference, Atlanta, 1974.

6. K. Shivanandan, J. R. Houck, and M. O. Harwit, Phys. Rev. Letters 21, 1460 (1968).

7. D. Muehlner and R. Weiss, Phys. Rev. Letters 24, 742 (1970).

8. A. G. Blair, et al., Phys. Rev. Letters 27, 1154 (1971).

9. D. Muehlner and R. Weiss, Phys. Rev. D 7, 326 (1973).

10. P. P. Woody, et al., Phys. Rev. Letters 34, 1036 (1975).

11. E. I. Robeson, et al., Nature 251, 591 (1974).

12. For example R. Carpenter, S. Gulkis, and T. Sato, Astrophys. J. 182, 261 (1973).

13. e.g. see S. Weinberg, Gravitation and Cosmology (Wiley, New York, 1972).

14. B. Corey, private communication (1976).

15. R. A. Muller, et al., private communication (1976).

16. P. Boynton and B. Partridge, private communication (1976).

17. Quarterly Progress Report No. 112, Research Laboratory of Electronics, MIT, p. 23 (1975).

18. D. K. Owens, R. Weiss, and D. Muehlner (to be published).

Work supported by NASA Grant 22-009-526

PART V

OBSERVATIONAL TECHNIQUES

BALLOON-BORNE TELESCOPES FOR FAR-INFRARED ASTRONOMY

William F. Hoffmann
Steward Observatory
University of Arizona
Tucson, Arizona 85721

I. INTRODUCTION

I would like to present some of the background and current activity in balloon-borne infrared astronomy, to examine a number of astronomical objectives in terms of their technical requirements, to describe some of the unique characteristics of ballooning techniques, and to illustrate how the objectives and techniques are mated with some specific examples.

Infrared astronomy from balloon platforms began in the early 1960's with two experiments: One of these was the Johns Hopkins' Gondola which was used for spectral measurements of Venus from 1.7 to 3.4 microns (Bottema, M., Plummer, W., and Strong, J., 1964, Ap.J., 139, 1021). The other early experiment was on a flight of Stratoscope II which carried a 0.8 to 3.1 micron spectrometer for stellar observations. This experiment provided an upper limit to the contribution of ice crystals to interstellar extinction (Woolf, N.J., Schwarzschild, M., and Rose, W.K., 1964, Ap.J., 140 833).

The first far-infrared balloon-borne observations (the measurement of 100 micron radiation from the galactic center) were made just 8 years ago (Hoffmann, W.F., and Frederick, C.L., 1969, Ap.J. (Letters), 155, L9). In the past 8 years, balloon-borne infrared telescopes have produced a variety of results including mapping of the galactic center region, mapping diffuse emission of the galactic plane, measurement of thermal radiation from interstellar dust clouds associated with HII regions, surveys of portions of the Milky Way, measurement of the submillimeter cosmic background flux, isotropy, and spectra, measurement of far infrared brightness of the sun, and measurement of solar spectral lines.

G. G. Fazio (ed.), Infrared and Submillimeter Astronomy, 155-168.

At the present time approximately thirty groups around the world are actively pursuing a variety of balloon-borne infrared observing programs. Table I is a list of groups which have flown an infrared astronomical balloon experiment during the 1970's. This list includes all experiments which have been flown whether they have succeeded or not. It does not include experiments under preparation. It contains only one entry for groups which have evolved more than one experiment.

Groups active in balloon-borne infrared astronomy are widely spread over the world. This list includes one group each in Argentina, Australia, Belgium, Italy, Japan, The Netherlands, and Switzerland. There are two groups each in France and Germany, three groups in England, and eleven in the United States. The largest portion of lauchings of scientific balloon experiments has been carried out from the National Scientific Balloon Facility in Palestine, Texas. Other launches are carried out in Argentina, Australia, France, and Japan.

There is also considerable variety in the astronomical goals. Seven experiments are devoted primarily to far infrared surveys, seven to measurements of the cosmic background radiation, four to high resolution mapping and photometry, one to measurement of diffuse emission in the near infrared, five to solar brightness, polarization, and spectra, and one to planetary spectra.

II. ASTRONOMICAL OBJECTIVES

The astronomical objectives representative of current and future efforts can be grouped into six categories which determine their technical requirements.

1) Surveys. The surveys are generally searches for discrete sources of infrared radiation over a portion of the sky. The survey telescopes must have a large beam for rapid coverage of the sky. This results in a large throughput (area x solid angle) of the telescope. For example, for a 40 cm telescope with a 12 minute of arc beam in the sky, the throughput is .012 cm^2 - steradian. To provide reasonable source identification, the positional accuracy of the surveys must be on the order of 1 arc minute. The survey requirements dictate a modest stability and a means of scanning the telescope. Most importantly, since the fundamental limit to the noise performance of an infrared detector is the thermal background radiation falling on the detector, with these large throughput survey telescopes it is important that the background from the sky and the telescope be low.

2) High Resolution Mapping. High resolution mapping dictates a small beam in the sky, usually close to the diffraction limit of the telescope. This results in a relatively low throughput

TABLE I
BALLOON-BORNE INFRARED ASTRONOMY
Flights 1970 - 1976

Ames Research Center	C. Swift	Pointed IR Telescope
Center for Astrophysics/ University of Arizona	G. Fazio, F.J. Low	High Resolution Far IR
Fraunhaufer Institute	Kiepenhauer	Near IR Solar Spectra
Geneva Observatory	E. Müller, F. Kneubühl	Far IR Solar Brightness
Goddard Space Flight Center	R. Hanel	IR Scanning Interferometer
Goddard Institute for Space Studies/ University of Arizona	W.F. Hoffmann, C. Frederick	Far IR Surveys and Mapping
Institute of Astronomy and Space Physics (Argentina)	E. Gandolfi	Far IR Survey
Leeds University	P. Marsden	3° Cosmic Background
Massachusetts Institute of Technology/ University of Arizona	W. Lewin, F.J. Low	Far IR Survey
Massachusetts Inst. of Technology	R. Weiss, D. Muehlner	3° Cosmic Background
Max Planck Institute, Heidelburg	D. Lemke, W. Hoffmann	Far IR Survey Near IR Photometry
Melbourne University	R. Thomas	Far IR Photometry
Nagoya University	S. Hayakawa K. Matsumoto	 2.4μ Diffuse Emission
National Center for Scientific Research (France)	G. Chanin	3° Cosmic Background
National Center for Scientific Research (France)	P. Grenier	3° Cosmic Background
Princeton University	P. Henry, D. Wilkinson	3° Cosmic Background
Queen Mary College, London	P. Clegg, J. Beckman	3° Cosmic Background
University of Arizona	F.J. Low, W. Poteet	Far IR Surveys and Mapping

TABLE I (Continued)

University of California, Berkeley	P. Richards, J. Mather	3° Cosmic Background
University College, London	R. Jennings	High Resolution Far IR
University of Florence	B. Melchiori	
	G. Dall'Oglio	Far IR Polarization
University of Groningen	R. J. van Duinen	
	H. Olthof	Far IR Photometry
University of Liege	R. Zander	
	L. Delbonille	IR Solar Spectra
University of Massachusetts	J. Strong	Solar Corona
Washington University	M. Friedlander	Far IR Survey

compared to the survey experiments and hence less critical dependence on very low background systems. For example, a diffraction limited 40 cm telescope operating at the wavelength of 100 microns would have a throughput of .0003 cm^2 - steradians (1/36 that of the same size survey instrument). The mapping experiments require scanning capability and substantially better stability than the survey experiments.

3) Faint Source Photometry and Spectroscopy. These experiments require a large aperture, a small beam to minimize the throughput, and accurate pointing with some means of offset guiding providing pointing errors of less than 10 seconds of arc. Because of the technical difficulties of realizing these requirements, faint source photometry and spectroscopy from balloon experiments is just beginning.

4) Cosmic Background Flux, Isotropy, and Spectra. Because the cosmic background provides a very low surface brightness, these experiments involve very large throughput systems - generally small telescopes with a relatively large beam providing a spatial resolution of about 7°. In addition, since the background flux is not only a source of detector noise, but is indistinguishable from the signal itself, these systems require a particularly low background for the instrument and the sky. As a consequence, the instrument itself is generally totally cryogenically cooled.

5) Diffuse Low Surface Brightness Mapping. This has very much the same requirements as the cosmic background experiments with the exception of a somewhat higher spatial resolution (approximately 1°).

6) Solar System Measurements. Because these objects are visually bright, spatially resolved, and have a high surface temperature, they do not place particularly difficult requirements on guiding or detector sensitivity. They generally do require low atmospheric contamination, high spectral resolution and accurate calibration.

III. SOME CHARACTERISTICS OF BALLOONING TECHNIQUES

The key to the value of balloon-borne observations for infrared astronomy lies in the behavior of atmospheric absorption by triatomic and diatomic molecules and by methane. Figure 1 shows the transmission of the atmosphere from 100 to 50 microns at 4 altitudes. (Traub, W. A., and Stier, M. T., 1976, Applied Optics, 15, 364) The transmission is given for an air mass of two. The altitudes are: 4.2 kilometers (13800') the height of the Mauna Kea Observatory, 14 kilometers (46,000') the approximate altitude of aircraft flights, 28 kilometers (92,000') the lower end of the

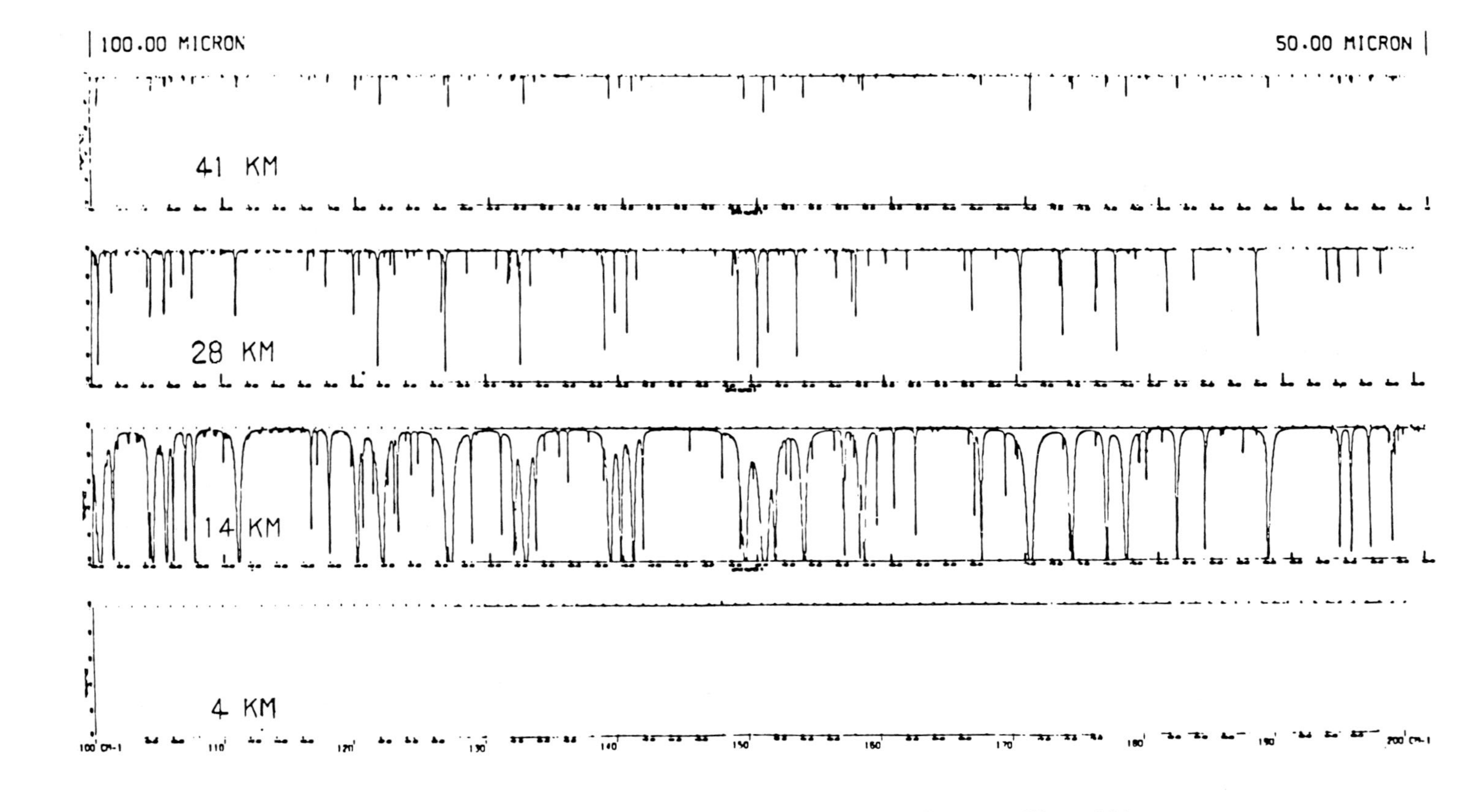

Figure 1 Transmission of the Atmosphere 50 - 100 microns

(Traub and Stier)

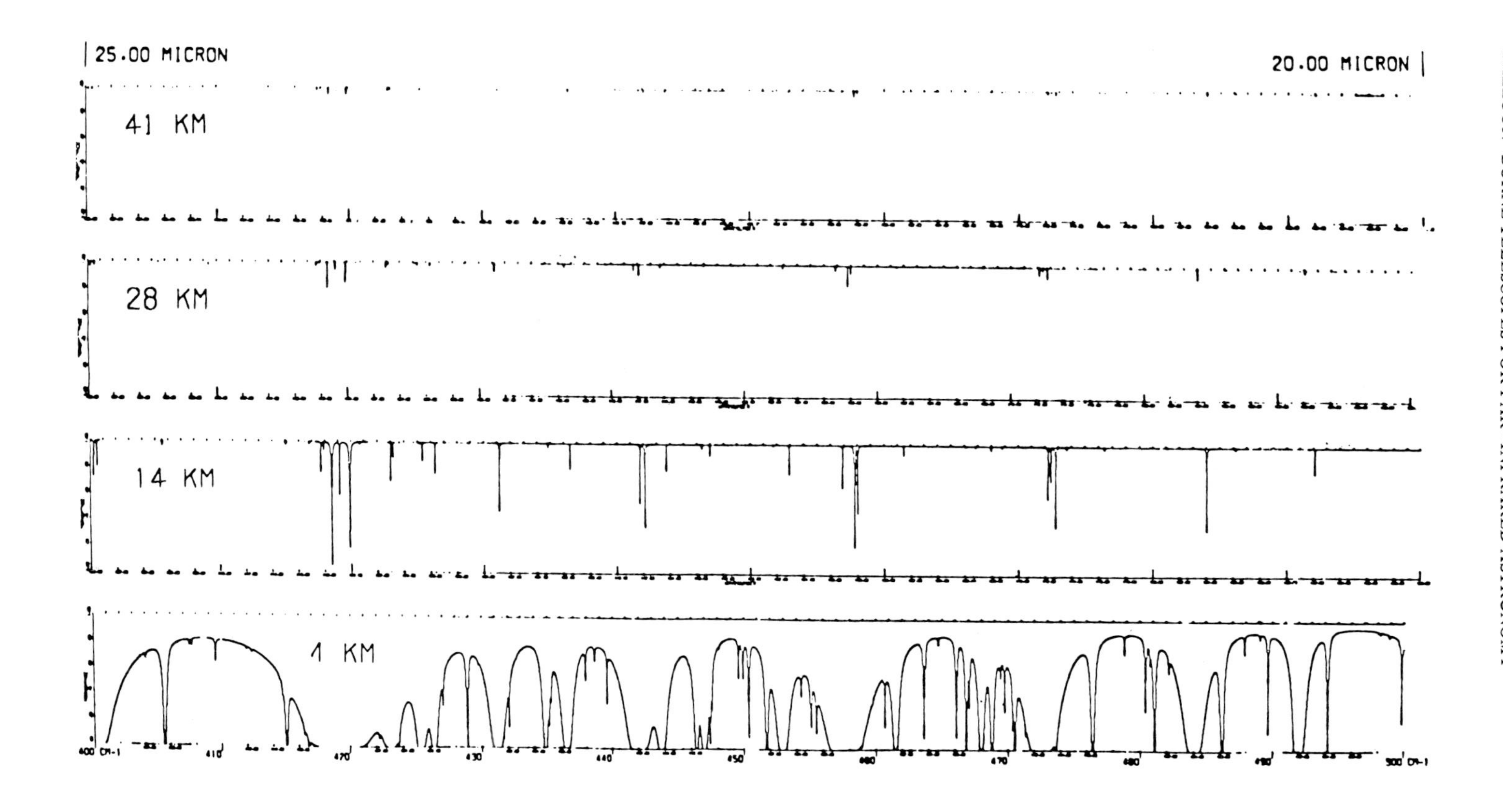

Figure 2 Transmission of the Atmosphere 20 - 25 microns

(Traub and Stier)

range of balloon altitudes, and 41 kilometers (135,000') the mid-range of balloon altitudes. It is striking to see the lack of any transmission at a mountain top observatory, the many very broad and generally saturated lines at aircraft altitudes, progressing to extremely sharp generally unsaturated lines at the lower balloon altitudes to rather faint lines at the high balloon altitudes. Figure 2 provides similar data for the spectral region 25 - 20 microns. This figure shows the very great improvement in the transmission at balloon altitudes over that through the atmospheric window at a high mountain observatory.

Figure 3 shows the integrated emissivity from 1 micron to 1 millimeter. From this figure it can be seen that the atmospheric emissivity at 100 microns is .03 at 28 kilometers, is .004 at 41 kilometers, and by interpolation is .015 at 34 kilometers (112,000'). At 22 microns where the background radiant flux provides a particularly significant limit to the detector sensitivity since this wavelength is close to the peak of the atmospheric Planck radiation curve, the emissivity is .007 at 28 kilometers, .0002 at 41 kilometers, and .00045 at 34 kilometers. These low emissivities at 22 microns provide a substantial gain for sky surveys in this region of the spectrum over attempts to carry out such surveys through the atmospheric window from the ground.

Now we will consider the means of reaching these altitudes: the balloon system. Currently used balloons provide a wide range of altitude and payload capability. Balloons for carrying 500 kilograms to 34 kilometers are modest in size, complexity and cost. Balloons for carrying this payload to 41 kilometer are commonplace although expensive. The record ballooning altitude is 52 kilometers (170,000'). The record weight is 5,000 kilograms.

Figure 4 shows a typical balloon train system as it would look when the balloon is fully extended at altitude. In this illustration the balloon is a 60,000 cubic meter (2 million cubic foot) balloon which can carry a 500 kilogram load to an altitude of 30 kilometers. The balloon diameter is 48 meters. The gondola is hung on a cable ladder connected to a parachute which is attached to the balloon with a fitting which is released on termination of the flight. The length of the suspension is 46 meters. In this configuration the balloon occults a portion of the sky down to a zenith angle of 20°. This represents 6% of the sky above the horizon and is of no consequence for surveys and photometric observations of discrete objects since the occultation of a given object is transient lasting at most 3 hours for a launch latitude of 30°. For cosmic background experiments it is desirable to observe far closer to the zenith to minimize to the air mass. It is not practical to increase the length of the suspension much beyond that shown in Figure 4 because of the increased difficulties of launch. However, a number of experiments

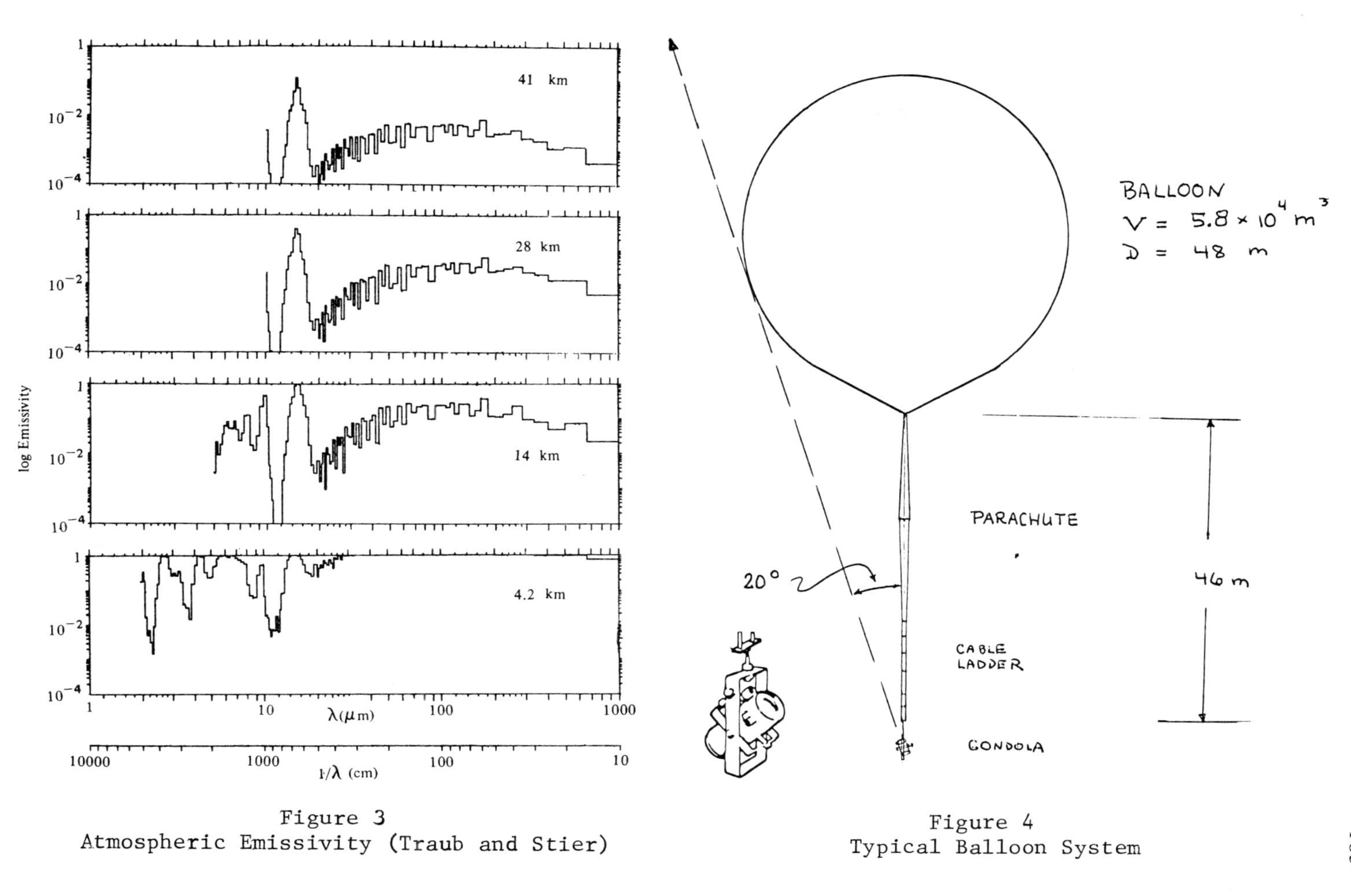

Figure 3
Atmospheric Emissivity (Traub and Stier)

Figure 4
Typical Balloon System

have flown with a reel containing 610 meters of nylon cord connected between the cable ladder and the gondola. After launch, this reel is unwound on command thereby increasing the length of the suspension. With this added length suspension the zenith angle occulted is 2 degrees.

The motions of the balloon system shown in Figure 4 place a number of technical requirements on a gondola stabilized for astronomical observation. The dominant motion of the system is the motion with the stratospheric winds which can be from 0 to greater than 100 miles an hour at float altitude. The wind speed rises very rapidly at altitudes above 30 kilometers. Typical wind speeds of thirty miles an hour at an altitude of 30 kilometers permit observing periods of 10 to 12 hours within a trajectory of 350 miles (the telemetry range of the National Scientific Balloon Facility in Texas).

The dominant motion that the stabilized gondola must be isolated from is a slow rotation of the entire balloon which reaches a maximum speed of one revolution per 8 minutes slowly changing and occasionally reversing during a flight. An additional motion is the pendulum motion of the entire suspension, gondola, and balloon with a period of approximately 15 seconds. The amplitude of this motion is generally less than or equal to 1 arc minute. An additional pendulum mode of the system is a rocking or double pendulum mode in which the gondola rocks back and forth about its center of mass. This mode typically has a frequency of 1 - 2 Hz and can easily be excited by the stabilization system. This is a particularly commonplace problem with gondolas whose azimuth stability is derived from a magnetometer which is sensitive to tilt as well as azimuth motion of the gondola. A number of means have been devised to overcome this problem including maintaining a large frequency separation between the azimuth stabilization system and the rocking frequency mode so that the stabilization loop has low gain at the rocking frequency, providing a mechanical decoupling and damping mechanism which inhibits the excitation of the rocking mode, and using a gyroscope instead of the magnetometer for the short term error signal determination.

Two axes, azimuth and elevation, are adequate to provide pointing of the telescope. For survey and other low accuracy systems, it is necessary to provide stabilization for only one of these axes, the azimuth axis, for which a magnetometer reference is often used. The elevation angle is then set relative to the gravity direction. For high resolution mapping, faint source photometry, or spectroscopy where greater pointing accuracy is required, stabilization of both axes are necessary in order to isolate the telescope from the pendulum motion of the gondola and suspension. These gondolas utilize gyroscopes and/or a star tracker as a reference. For these systems the two axis alt-azimuth system is not completely satisfactory because these coordinates are not orthogonal and the azimuth stabilization system must move the entire inertia of the gondola. A preferable system is shown in Figure 4. This has 3 axes: azimuth, elevation, and cross elevation. The precision stabilization is done entirely in elevation and cross elevation.

IV. ILLUSTRATIONS

Figure 5 shows the gondola of the Goddard Institute for Space Studies/University of Arizona which has done a dozen flights since 1969. The telescope is a 30 cm f/5 Newtonian with a 12 minute of arc beam. It operates with a solenoid driven secondary chopper operating at 20 Hz. The gondola is stabilized in azimuth to 1 - 3 arc minutes utilizing a flux-gate magnetometer. The elevation angle is determined relative to gravity. The gondola can be oriented to any elevation and azimuth. This instrument is used only for mapping and surveying. It operates with a scan motion in elevation or azimuth up to 10° at a rate of .3 degrees per second. The gondola is 4 meters tall and weighs 300 kilograms. The achieved rms flux equivalent noise is 3,000 $Jy/\sqrt{Hz}$.

A new 40 cm helium cooled telescope is presently being prepared for flight in a collaboration of the University of Arizona and Cornell University. In this new system the optics is maintained at a temperature of 4 to $10^{o}K$ and the detector at $1.8^{o}K$. Air is prevented from entering the telescope where it would condense on the cold optics by a four micron thick polyethylene membrane. This system is being prepared for shipment to Palestine, Texas for launch in late June 1976. The potential sensitivity of this instrument flying at an altitude of 110,000 feet is rms flux equivalent noise of 34 $Jy/\sqrt{Hz}$, 20 microns, 49 at 50 microns and 62 and 100 microns.

Figure 6 shows the balloon gondola of University College, London, constructed by the Physics Engineering Group. This carries a 39 cm aperature Dall Kirkham telescope with a folding mirror which passes the telescope beam out through the elevation axis. This telescope has operated over a band from 40 microns to 350 microns with a 3.5 minute of arc beam. It is a two axis system in which both the azimuth and elevation axes are stabilized relative to a star tracker with a sensitivity of fourth magnitude and an offset capability of ±5 degrees. The stabilization accuracy is ±30 arc seconds. The gondola weighs 500 kilograms. The University College group is now preparing a new advanced gondola with three axes of stabilization (elevation, azimuth, and cross-elevation) in order to provide more precise pointing. In collaboration with ESTEC they are preparing a Michelson interferometer for high resolution spectroscopy.

Figure 6 University College Gondola

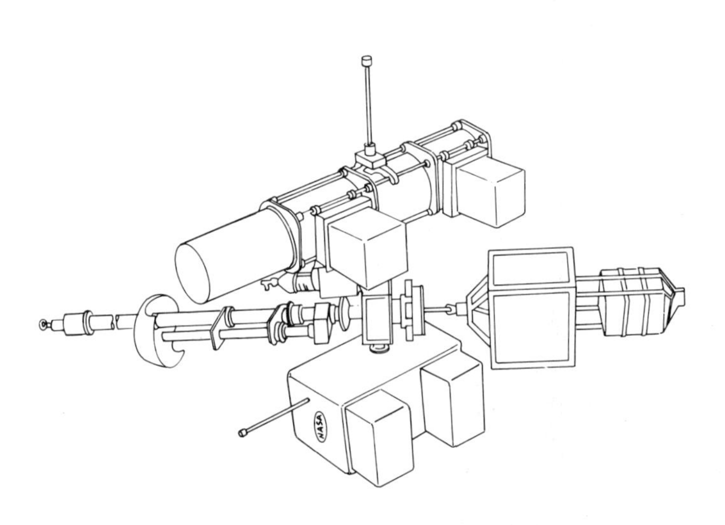

Figure 5 Goddard Institute for Space Studies / U. of Arizona Gondola

Figure 7 shows the one meter telescope of the Center for Astrophysics (Smithsonian Astrophysical Observatory - Harvard College Observatory)/ University of Arizona. This is a two axes system with both axes stabilized relative to gyroscopes. This system also has a television monitor for field acquisition and guiding. It operates over a wavelength band 40 - 250 microns and also narrower bands at 20, 40, 80 and 120 microns. It has achieved equivalent flux noise power at 100 microns of $70 \ \mathrm{Jy}/\sqrt{\mathrm{Hz}}$. The resolution is 30 arc seconds. This gondola weighs 1,800 kilograms.

Figure 8 shows the gondola of the University of Groningen. The basic gondola has been designed by Ball Brothers providing altazimuth stabilized system utilizing a star tracker with an offset field of 8° and pointing stability of ± arc minute. It carries a 60 cm telescope and weighs 500 kilograms. It operates with four color far infrared photometry utilizing reimaging and restrahlung filters.

The fifth illustration of a gondola system is the low emissivity telescope system of the University of Arizona reported

Fig. 7. Center for Astrophysics (HCO/SAO) and University of Arizona Gondola.

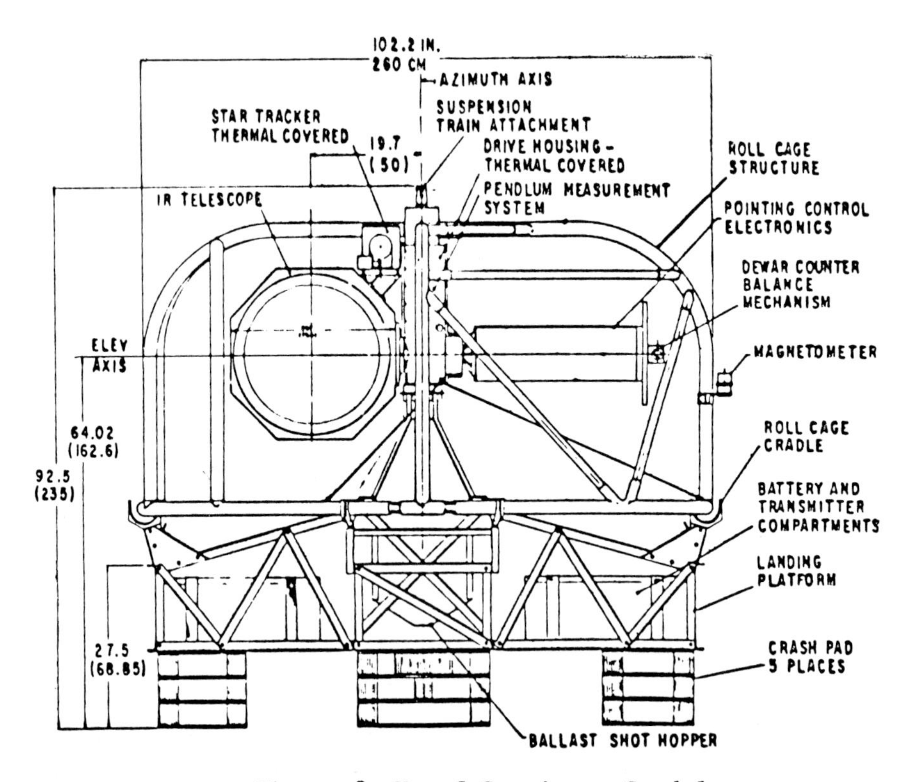

Figure 8 U. of Groningen Gondola

later in this session. It is a 3 axes system with azimuth and approximate right ascension and declination. It is stabilized only in azimuth relative to a magnetometer gravity. The unique feature of this telescope is the very low emissivity optical design and gold plated surfaces to provide very low background for high throughput surveying and mapping.

V. SUMMARY

Ballooning offers unique capability for far infrared astronomy due to very low atmospheric line contamination and low infrared thermal flux background. There are a wide variety of astronomical goals that can take advantage of these characteristics of the stratosphere including absolute flux measurements and surveying and mapping of discrete sources, which utilize high throughput systems and high resolution mapping, faint source photometry, and spectroscopy which utilize highly stabilized pointing systems. The wide variety of infrared groups and instruments reflects the many different technical approaches which can be taken to realize these astronomical goals. A particular value of ballooning as a technique in far infrared astronomy is its great flexibility and relative speed and economy in responding to new ideas.

INFRARED OBSERVATIONS FROM AN AIRBORNE PLATFORM

D.A. Harper

Yerkes Observatory, University of Chicago,
Williams Bay, Wisconsin, USA.

ABSTRACT

The National Aeronautics and Space Administration's Gerard P. Kuiper Airborne Observatory has now been operational for approximately one year. During that time it has provided ~ 550 hours of observing time at altitudes of 13-14 km to 14 separate research groups. The experiments flown include photometers and spectrometers designed for observations in spectral bands ranging from 1-1000μ.

The Observatory consists of a gyrostabilized 0.91 m reflecting telescope mounted in a C-141 aircraft. The basic stability of the telescope is better than ±2 " and an absolute pointing accuracy of greater than ±5 " can be achieved. The guidance system can easily be programmed for extended observations of several hours duration at a fixed celestial position or for various types of raster scans covering areas of $\lesssim$1 square degree.

Instrumentation and observing procedures for use with the Observatory are closely analogous to those used for ground-based infrared observations through atmospheric "windows." Problems associated with atmospheric absorption, instrumental and sky background, and noise associated with atmospheric turbulence are also similar to those encountered at terrestrial sites - although they are typically much smaller in magnitude. The photometric sensitivity achieved in the far infrared is comparable to that attained by balloon-borne experiments not specifically designed for extremely low instrumental backgrounds. Although the ultimate sensitivity of the present aircraft-mounted cannot be

G. G. Fazio (ed.), Infrared and Submillimeter Astronomy, 169-170.

expected to be competitive with future generations of low-background balloon or satellite-borne instruments, its accurate pointing, relatively large aperture, and operational flexibility have permitted the initiation of a broad range of scientific investigations which are of high current interest and which can provide sound observational and technological data to aid in the design of more advanced instrumentation.

SPECTRAL AND POLARIMETRIC INSTRUMENTATION FOR THE AIRCRAFT ASTROPHYSICAL INVESTIGATIONS IN THE RANGE 50-500 mkm

G.B. Sholomitski, V.A. Soglasnova, I.A. Maslov,
V.D. Gromov, M.S. Khokholov, V.V. Artamonov
Space Research Institute, Academy of Sciences, USSR

ABSTRACT

Receiving equipment for aircraft astrophysical investigations in the far infrared has been developed and flown in 1974-1975 on board the AN-30 aircraft. Four types of infrared and submillimeter detectors mainly of photoresistor type are used in the wavelength region from 10 mkm to 1 mm. Thermal background at the detectors is limited by cooled bandpass filters and by optimal matching of the detectors with the telescope optics. The spectral filters used are combinations of quasiresonance metal mesh filters of different structure and Q-factor from 2 to 7 with the Yamada cut-off powder filters. A tunable Fabry-Perot interferometer, a polarimeter for linear polarization measurements and an aircraft 25 cm-telescope are briefly described.

G. G. Fazio (ed.), Infrared and Submillimeter Astronomy, 171.

FAR-INFRARED OBSERVATIONS WITH A SMALL, LOW-BACKGROUND, BALLOON-BORNE TELESCOPE

Frank J. Low, Wade M. Poteet, and Robert F. Kurtz

Lunar and Planetary Laboratory and Steward Observatory
University of Arizona, Tucson, Arizona

ABSTRACT

A single-mirror, off-axis telescope 20cm in diameter has been developed for observations in the 50-300μ band. Because of its low background properties, a large field of view and high sensitivity are obtained. With a 15 arcminute field of view the peak-to-peak noise equivalent flux with one second integration time is approximately 1000Jy. The performance of this telescope and its associated three-axis pointing system will be discussed and the results of a limited far-infrared sky survey will be presented.

G. G. Fazio (ed.), Infrared and Submillimeter Astronomy, 172.

SIMULTANEOUS MULTI-COLOR FAR-INFRARED PHOTOMETRY

I. Gatley, E.E. Becklin, M. Werner

California Institute of Technology

ABSTRACT

An astronomical experiment is described which allows simultaneous viewing of the same field of view in well defined bands at 30, 50 and 100μm with $\Delta\lambda/\lambda \sim {}^1/_2$. Color temperatures as well as surface brightnesses are measured directly; the system has obvious advantages for airborne and balloon platforms where good absolute pointing accuracy and reproducibility are difficult. This experiment has been flown successfully on the NASA 91 cm airborne telescope.

G. G. Fazio (ed.), Infrared and Submillimeter Astronomy, 173.

COMPARISON OF PHOTOCONDUCTIVE-BOLOMETER DETECTORS ON AN AIRBORNE SYSTEM

K. Shivanandan, D.P. McNutt, W.J. Moore

Naval Research Laboratory, Washington, D.C., USA

ABSTRACT

A Ge:Ga photoconductive detector at 100μm and a GaAs epitaxial photoconductive detector at 285μm has been optimized for sensitivity comparison with a conventional bolometer system of Yerkes Observatory. The response of the photoconductive detectors were studied in the laboratory under two background conditions: baffles and Fabry Optics. A direct comparison of the photoconductive detectors to the bolometer will be made on the 91.5 cm telescope aboard the NASA C-141 Airborne Infrared Observatory.

G. G. Fazio (ed.), Infrared and Submillimeter Astronomy, 174.

PART VI

O B S E R V A T I O N A L   T E C H N I Q U E S

INFRARED ASTRONOMICAL SATELLITE (IRAS)

R.J. van Duinen

University of Groningen, Kapteyn Astronomical Institute,
Space Research Department, Groningen, The Netherlands.

1. INTRODUCTION

In this paper we will describe the scientific objectives and conceptual design of the Infrared Astronomical Satellite: a joint proposal for an all sky infrared survey between the Netherlands, the United Kingdom and the United States. This paper is based on the results of a series of studies that have been made recently. Initially the proposal for an infrared survey satellite was made to the Dutch government on the occasion of the launch of the ANS satellite in August 1974. This proposal formed the basis of a study in the Netherlands during 1975. In the course of the study coordination with similar proposals in the United States resulted in the creation of a joint Netherlands-UK-US mission definition team. This joint team concluded its work in May 1976. As a large number of individuals and institutes contributed to the study report and my presentation heavily draws on it, this paper should be considered as a summary of the work performed by the joint study team.

2. SCIENTIFIC OBJECTIVES (expected performance)

Perhaps the single-most important aspect of an infrared survey is the sensitivity limit to which the survey should yield reliable results. Since ground-based work in the atmospheric windows can reach the 6th magnitude around 10μm it seemed important to achieve a survey limit of at least that value in the 8 to 30μm range. At longer wavelengths a "natural" limit may be defined by a tie-in with present day high frequency radio surveys.

G. G. Fazio (ed.), Infrared and Submillimeter Astronomy, 177-183.

The accuracy with which (point) sources can be located on the sky should allow follow-up studies and identification. An accuracy of approximately one arcminute was selected as a compromise between the number of detectors in the array, the repetition of the survey (related to the array width), the data storage and handling problem and also - for the longer wavelength channels - the size of the telescope.

The performance limits in terms of sensitivity and positional accuracy are summarized in table 1. As it turns out, the sensitivity in the two short wavelength channels is limited by infrared emission of the zodiacal dust. In other words the proposed short wavelength detectors are expected to operate under background limited conditions where the background source is zodiacal emission and not - as usually is the case - emission from the telescope and the atmosphere. Here the full potential of observing from space is used: the sensitivity gain is achieved by operating the detectors at the ultimate limit of performance by cooling the telescope to cryogenic temperatures and by avoidance of atmospheric emission. In the two longer wavelength channels the expected performance is based on detectors that are not - at this time - readily available. A detector development program has been initiated with the industry. It is very well possible that the quoted performance will be surpassed in which case the sensitivity at longer wavelengths will be even more exciting. It is believed that the quoted values for the detectors can be obtained with reasonable efforts in improvement of detector fabrication techniques.

At the given levels of performance the sensitivity of the survey can be judged by quoting distances to which some typical infrared objects may be observed. In table 2 a representative summary is given. For galactic research it is of interest to note that typical protostellar objects can be seen out to distances of 30Kpc. One may, therefore, expect to achieve a wealth of data relevant to statistical studies of star formation. At the same time one should also note that at the highest sensitivities considerable source confusion will occur in the galactic plane due to the profusion of stellar sources. Consequently, it may be impossible to exploit the full sensitivity in the final analysis of the survey data obtained in the plane.

Extra galactic sources like M82 may be seen out to 70Mpc, thus beyond the local group. The volume of space sampled for active galaxies is considerable. One may speculate about detection of "infrared" galaxies; when they exist, there is a good chance that IRAS will discover a reasonably large sample of them.

Table 1

Channel	Detector material	No	NEFD (W/cm^2)	NESD(Jy)		Dimensions (arcmin.)	Positional error(arcmin	
				ultimate	expected		in scan	cross scan
8-15μm	Si:As	19	$1.1 x 10^{-19}$	0.007	0.007	1.2 x 2.4	0.4	0.7
15-30μm	Si:Sb	19	$8.3 x 10^{-20}$	0.010	0.010	1.2 x 2.4	0.4	0.7
30-60μm	Ge:Be	19	$4.7 x 10^{-20}$	0.011	0.1	1.2 x 2.4	0.4	0.7
60-120μm	Ge:Ga	10	$4.2 x 10^{-20}$	0.018	0.2	1.8 x 4.8	0.5	1.4

This table summarizes some characteristics and the expected performance of IRAS. The noise equivalent flux density and noise equivalent spectral density have been computed on the basis of a zodiacal emission of 3×10^{-11} W cm^{-2} sr^{-1} μ^{-1} and a telescope temperature of 16K, while the orbital scan rate of 3.5 arcmin/sec is used to estimate the dwell time. Note, that the more conservative estimate for the sensitivity of the long wavelength channels is used in Table 2.

Table 2

Object	Class	Actual Distance	Max. dist. detectable at λ with baseline system		
			10 μ (8-15 μ) .06 Jy†	20 μ (15-30 μ) .1 Jy†	100 μ (75-125 μ) 2 Jy†
Solar System					
Asteroid at 300K	3 km diameter		1 AU		
Galactic Sources					
α Lyr	AO V star	8 pc	180 pc**	56 pc	
α Ori	M supergiant	.2 kpc	90 kpc**	25 kpc	
o Cet	Mira variable	.04 kpc	8 kpc**	5 kpc	
IRC + 10216	Carbon star large in IR excess	.3 kpc	300 kpc**	32 kpc	
BN	"protostar"	.5 kpc	35 kpc	30 kpc	
KL neb. in Orion	"IR star cluster" large silicate absorption	.5 kpc		100 kpc	
NGC 7027	planetary nebular bright in IR	1.7 kpc	80 kpc		
Trapezium	Typical H II region-no silicate absorption	.5 kpc	50 kpc	50 kpc	
W51-IRS 1	Giant H II region/mole cloud large silicate absorption	.6 kpc	240 kpc	580 kpc	1600 kpc (all of W51)
W51-IRS 2	Giant H II region/mole cloud mod. silicate absorption	.6 kpc	410 kpc	750 kpc	
M17	Giant H II region. Bright near IR	1.6 kpc	940 kpc	1700 kpc	600 kpc
OMC 2	Molecular cloud w/imbedded near IR sources	.5 kpc	15 kpc	13 kpc	20 kpc
ρ Oph Dark Cloud	Dark cloud/mole cloud	.15 kpc			15 kpc
Sgr B2	Giant H II/mole cloud	10 kpc			5 Mpc
Galactic Center		10 kpc	4 Mpc	2.5 Mpc	12 Mpc
Extragalactic Sources					
NGC 253	Spiral galaxy. Some nuclear activity	2.4 Mpc	35 Mpc	60 Mpc	70 Mpc
M82	Dust galaxy. Some evidence for explosive activity	3 Mpc	50 Mpc	70 Mpc	80 Mpc
NGC 1068	Seyfert galaxy	22 Mpc	340 Mpc	400 Mpc	230 Mpc
Mk 231	Markarian galaxy	230 Mpc	1100 Mpc		
3 C 273	Quasi-stellar object-variable	950 Mpc* v/c = .16	3300 Mpc* v/c = 0.5	3300 Mpc	

† limit for detection with signal-to-noise of 10:1 on proposed survey.
* assuming redshift is cosmological. ** assuming no interstellar 10μm absorption.

3. CONCEPTUAL DESIGN

The survey instrument consists of a cold 60 cm Cassegrain telescope equipped with a series of photoconductor arrays. The arrays are arranged in four wavelength bands: 8-15μm, 15-30μm, 30-60μm and 60-120μm. The sizes correspond to 1.2 x 2.4 arcminute in the 10, 20 and 50μ bands and to 1.8 x 3.6 arcminute in the 100μ band. The total width of the survey arrays is 30 arcminutes. In addition to the survey arrays the focal plane contains a few long wavelength detectors to extent the survey sensitivity to longer wavelength. Also, provisions have made for the incorporation of narrow band photometry. The latter will provide additional information which will be a powerful aid to source classifications e.g. through observation of the depth of the 9.7μm silicate feature along with the survey.
Signal modulation can be provided by the orbital scan motion. Continued attention will be given to the option to employ beam switching. Such an approach would be less susceptible to low frequency noise and would allow observation of weak sources by integration while pointing the satellite. Disadvantages are added complexity and possible encumbrance of the electronic suppression of the effects of energetic particles. The cryogen needed to cool the telescope and detector assemblies will be liquid Helium either in the superfluid state or partly superfluid and partly supercritical. Both solutions have been studied and seem entirely feasible.

Interference of ionizing radiation with infrared detectors is a potential source of problems. However, use can be made of the large difference in the charge deposit rate and therefore the rise time of the pulse caused by an ionizing particle as compared to the response to an infrared source. Using electronic pulse shape discrimination techniques it will be possible to limit the disturbance caused by particle hits to a small fraction of the total observing time.

Another area of concern is the occurrence of spontaneous noise spikes and the presence of man made and natural debris around the earth and in the solar system. Both phenomena could potentially affect the reliability of the survey since they are indistinguishable from astronomical infrared objects. For this reason there is considerable emphasis on repetition of the survey. In the survey instrument coincidence of source detections will be required. Moreover, scans in consecutive orbits will have a fifty percent overlap.
Then, the scanned area of sky will be re-observed about 20 days after the first (two) scans and a third time 40 days after the first scan. Finally, the second half year of operation will allow either selective or complete repetition of the survey.
It is felt very strongly that the numerous repetitions are needed to eliminate spurious detections, to discriminate against solar

system and earth orbiting objects and finally - very importantly- to detect variability in infrared sources. The satellite will be launched in a similar orbit as ANS: a sunsynchronous high inclination orbit. The altitude for IRAS is 900 km, a compromise between contamination by cryo-deposits and radiation interference. The survey is performed by scanning the sky at the orbital rate in small circles in planes which are perpendicular to the sun satellite axis(z-axis). The cone angle between the experiment viewing direction and the z-axis can be selected between 60 degrees and 120 degrees (provided no interference with the location of the earth horizon occurs). Experiment control and attitude control are performed by a fully programmable onboard computer (a similar concept was used in ANS). Apart from scanning, the satellite is also capable to perform attitude functions such as pointing in a preselected direction, raster scanning and fast or slow scan.

Every 12 hours a complete satellite observation program will be transmitted to the spacecraft from the control centre located in England and the data gathered during the previous 12 hours - which are stored on a high capacity tape recorder - will be transmitted to the ground. On the ground the total dump of about 5×10^8 bits will be put on tape and sent to the US for final processing. Also, quick look data will be generated at the control centre using simple algorithms for extraction of bright point-like sources. On the basis of both the scientific quick look and the engineering quick look the performance of the survey instrument and spacecraft will be monitored. Also, as required, adjustment of the observation programs can be initiated.

The final results of the survey can only be obtained after careful processing of the raw data. This includes the removal of spurious signals, the computation of source positions on the sky from orbit and satellite sensor data, the confirmation of source detections. Since a large number of detections in the 10μm channel will be of stellar origin it appears necessary to attempt identifications with star catalogues. Similarly one might consider identifications with HII regions and galaxies, and other categories of astronomical objects.

An important consideration is the form of the final database that will be made available to the astronomical community. If the user has access to a computing system perhaps it would be most appropriate to obtain a tape copy of the position ordered final catalogue file. Ideally such tape copies should be made available in a variety of formats(to suit various computer configurations) together with a software package to provide access to the data. The user could generate outputs in accordance with his particular needs. Other categories of users can be best supplied with partial print-outs in accordance with his wishes, or with overlays on Palomar sky prints. Conceivably, microfilm copies of a complete listing can also be provided.

It seems appropriate to address the time span between the start and the completion of the survey including the rather substantial processing of the raw data. The required high reliability of the survey dictates a total time span of at least two years. Obviously, on interesting discoveries follow-up studies should be initiated earlier. Such could be achieved best by an open and quick dissemination of preliminary new results through an appropriate channel each time the science team has agreed that there is reasonable confidence that the discovery is a real one and will satisfy all the criteria of the final processing. Such a procedure would also provide the data which are needed for selective confirmation of discoveries by IRAS itself during the second half year of operation.

The final decisions to go ahead with the IRAS project have not yet been made. Under a joint responsibility the Netherlands would provide the spacecraft, the United Kingdom would provide operational facilities and support while the United States provide the survey instrument, the launcher and launch support. Additional experiments such as a low resolution spectrometer and the long wavelength detectors are presently considered for inclusion in the focal plane by the Dutch and British groups.

USE OF THE LARGE SPACE TELESCOPE FOR INFRARED OBSERVATIONS

Gerry Neugebauer

Physics Department, California Institute of Technology
Pasadena, California 91125

It is appropriate that a discussion of the infrared capabilities of the (L)ST follows one about the IRAS since the (L)ST bears much the same symbiotic relationship to IRAS as the 200-inch Hale Telescope does to the 48-inch telescope which performed the complete optical survey of the sky.

The (L)ST is an observatory and its infrared capability is one facet of a complete astronomical observatory. A conceptual design of the entire spacecraft is presented in Figure 1 which shows that 4 axial bays and one radial bay into which instruments can be placed are available. Some relevant characteristics of the (L)ST spacecraft are listed in Table 1. The spacecraft, with a projected lifetime in excess of 15 years, will be launched from the shuttle, and will be maintained by visits from suited astronauts; modular construction is embodied throughout.

For the last several years, definition studies have been carried out to show the feasibility of seven candidate instruments; a brief tabulation of these taken from several (L)ST reports is given in Table 2. At the present time the decision as to which instruments will fly on the first flight is under discussion based on the limited space and budget available. It is encouraging that the working group made up of the team leaders of the definition teams plus several "at large" astronomers has included the infrared photometer among the first generation to be flown in the early 1980's.

The definition team for the infrared photometer consists of R. Hall, D. Kleinmann and G. Neugebauer. As a first generation experiment the team proposes a simple filter wheel photometer which works from 1 μ to 1 mm. A spectrometer with a spectral resolution of $\lambda/\Delta\lambda \sim 3 \times 10^4$ is proposed for a second generation experiment. The choice of a photometer as a predecessor of the

G. G. Fazio (ed.), Infrared and Submillimeter Astronomy, 185-194.

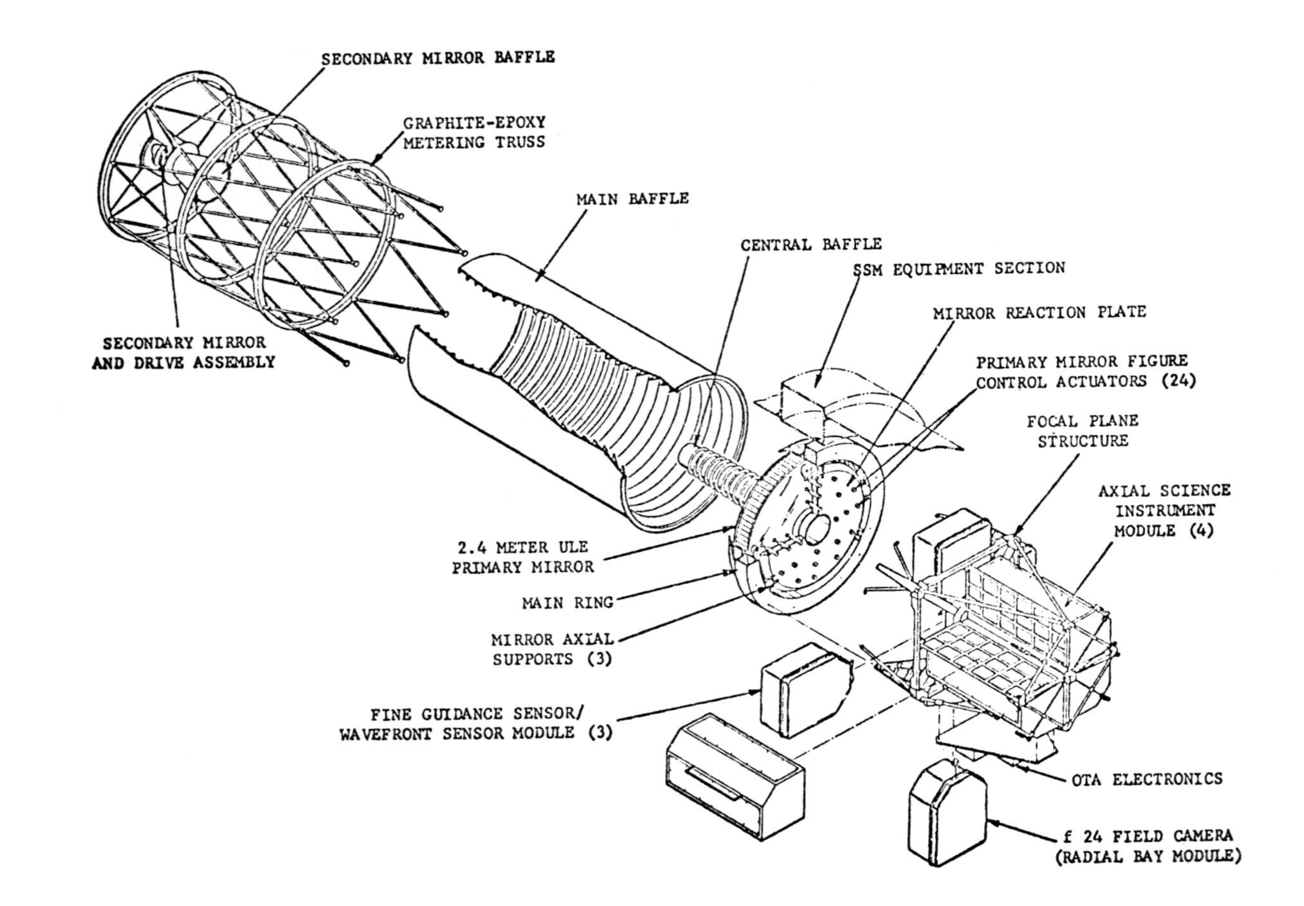

Figure 1 - An exploded view of one concept of the (L)ST. Scientific instruments are contained in four axial bay modules and a radial bay module.

TABLE 1

RELEVANT CHARACTERISTICS OF THE (L)ST SPACECRAFT

MIRROR:	2.4 m diameter f/2.2 ULE
OPTICS:	Ritchey Chrétien f/24 system
MIRROR TEMPERATURE:	294 ± 1 K
FIELD DIAMETER:	~20'
CENTRAL OBSCURATION:	31%
FIGURE:	0.05λ rms at 633 nm
SPACECRAFT ALTITUDE:	500 km
LIFETIME:	15 years with refurbishment
STABILITY:	0.006"
OBSERVATION TIMES:	up to 10 hours

TABLE 2

INSTRUMENTS DISCUSSED DURING DEFINITION PHASE

F24 Camera: A wide field camera of high spatial resolution together with appropriate isolating filters. Guidelined to use the SEC integrating television detector, it would be used for specific observations and for field mapping.

Faint Object Camera: A narrow field of view camera able to utilize the full resolution capabilities of the LST. Provided with a photon counting detector, it would be able to reach the faintest possible stars and galaxies.

Faint Object Spectrograph: A low resolution ($\lambda/\Delta\lambda \simeq 10^3$) spectrograph optimized for ultraviolet studies at spatial resolutions approaching 0.1 arc sec. Ultra low (100) and high (20,000) resolution modes allow a wide variety of problems to be attacked.

High Resolution Spectrograph: A versatile spectrograph primarily designed to give UV spectra with a spectral resolution on the order of 3×10^4.

Astrometry: By appropriate calibration and design the LST fine guidance system can be used to provide fundamental astrometric data of one full order of magnitude greater accuracy than can now be done.

Infrared Photometer: A straightforward two channel filter photometer capable of significantly higher spatial resolution than large ground telescopes when used at short wavelengths and high resolution and faint threshold limit when used at long wavelengths where the atmosphere is opaque.

High-Speed Area Photometer: A simple and reliable photometer which measures fluxes photometrically from the far UV (125 nm) through the visible (650 nm). The photometer incorporates high light throughout, wide spectral range, polarization sensitivity and time resolution.

spectrometer was dictated by the relative simplicity and reliability of a photometer, by the fact that it could be built with state of art components, and, finally, by the fact that it provides a logical scientific progression from simpler to more sophisticated problems in much the same way that photometry has preceded spectroscopy in ground-based infrared astronomy.

A schematic outline of a photometer for the (L)ST is provided in Figure 2; it has been drawn by one of the industrial teams determining the feasibility of proposed instruments. The photometer is designed to work at wavelengths from 1 μ, where the Airy disk of the (L)ST is 0.2", to 1 mm, where the disk is ~3'; obviously, it is designed to occupy one of the axial bays. Perhaps the most conspicuous feature of the design lies in the dewar of liquid He which will maintain the detectors at a temperature less than 2°K. This distinguishes the infrared experiment from the other proposed instruments in that it restricts the lifetime to a minimum of one year. The dewar provides the main untested component of the system but seems to be well within the capability of cryogenic technology.

With the exception of the dewar, the rest of the photometer is made of state of the art components. Chopping to reduce offsets due to the background radiation is provided by a focal plane chopper which oscillates an image of the primary mirror; throws from 0.4" to 210" at frequencies up to 50 Hz are planned. The photometer will have at least two detectors, a photoconductor such as Si:As or Si:P to cover wavelengths less than ~30 μ and a composite bolometer for the long wavelengths. Wavelengths will be isolated by a set of ~60 filters whose bandwidths will be chosen as appropriate to the wavelength; typically 10% spectral resolutions are planned. Apertures going from ~0.5" to 210" in steps of ~ $\sqrt{2}$ will be provided.

What kind of science will be stressed for the infrared photometer? In common with all the proposed instruments, the main advantage of the (L)ST lies in its high spatial resolution. At all wavelengths the spatial resolving power of the 2.4 m disk is important in its own right, and this provides the main thrust for the usage at the shorter infrared wavelengths. Thus it will be possible to map the Galactic center at 1 μ with a 0.2" aperture and comparisons of H II regions can be made at wavelengths up to 100 μ on a scale of 10". More importantly, the stability of the (L)ST allows one to determine structure on selected bright objects at a scale well below the diffraction limit by comparing scans across standard stars and across objects of interest. Thus, for example, we will be able to study the structure of the nuclear region of Seyfert galaxies at a scale set by the stability of the spacecraft (< 0.01") since the response profile of the photometer does not depend on time or "seeing" but is constant and reproducible. We have shown the viability of this procedure in studies of NGC 1068 using the 200-inch telescope but the time of sufficiently steady and good seeing on the ground is measured in hours

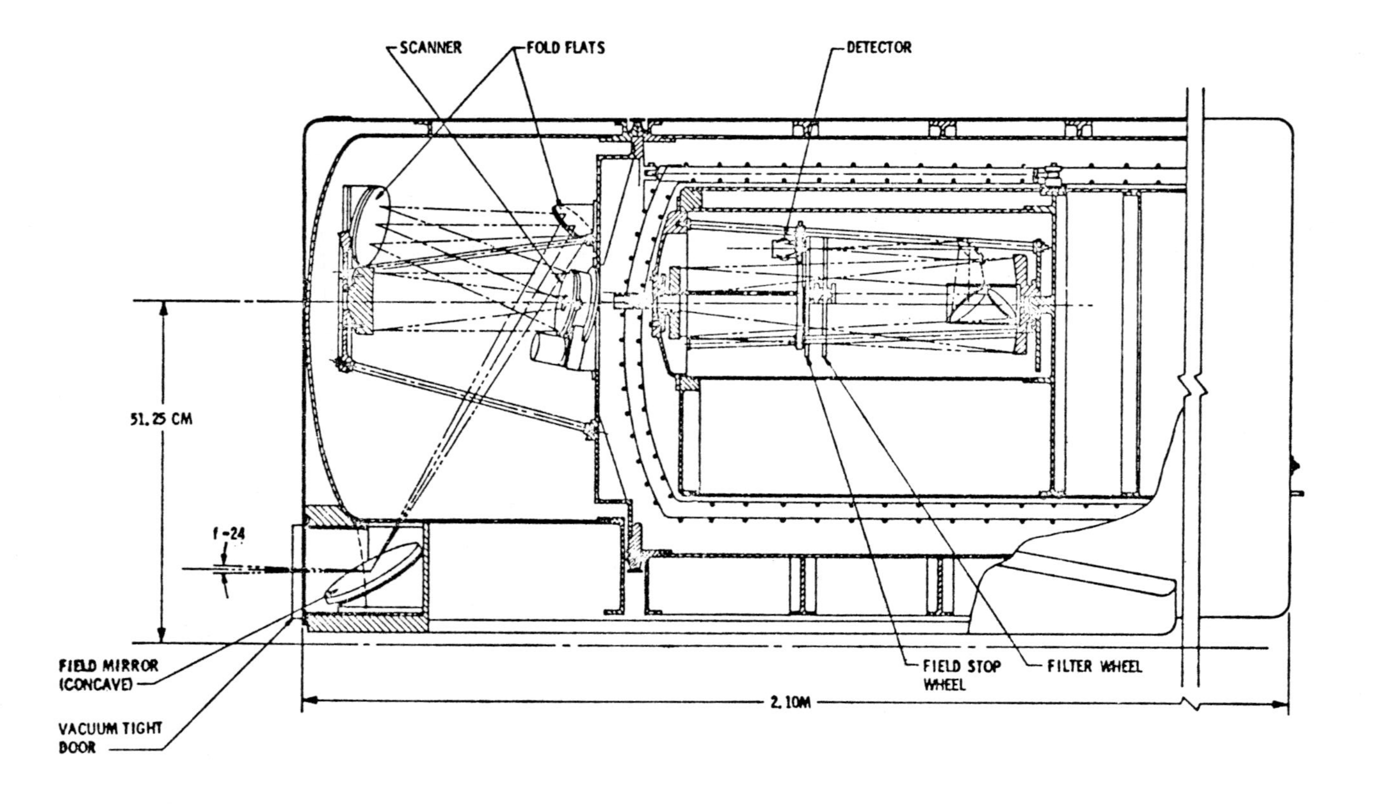

Figure 2 - One concept of the infrared radiometer layout is shown. The photometer contains a dewar with a one-year lifetime, the chopper for removing the background offset, two detectors and a set of variable apertures and filters for each.

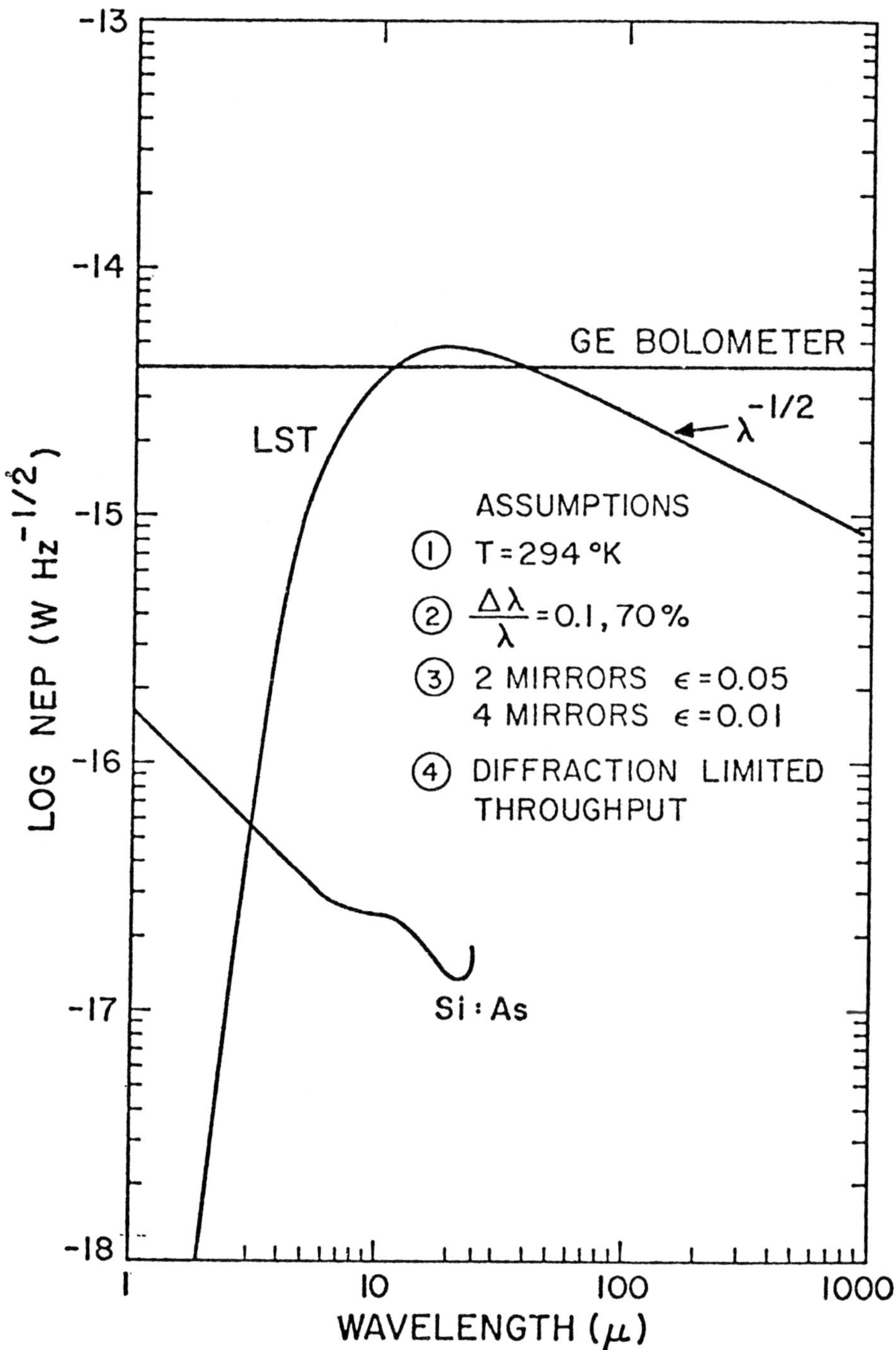

Figure 3 - The noise equivalent power produced by the fluctuations in the warm mirror of the (L)ST are shown as a function of wavelength.

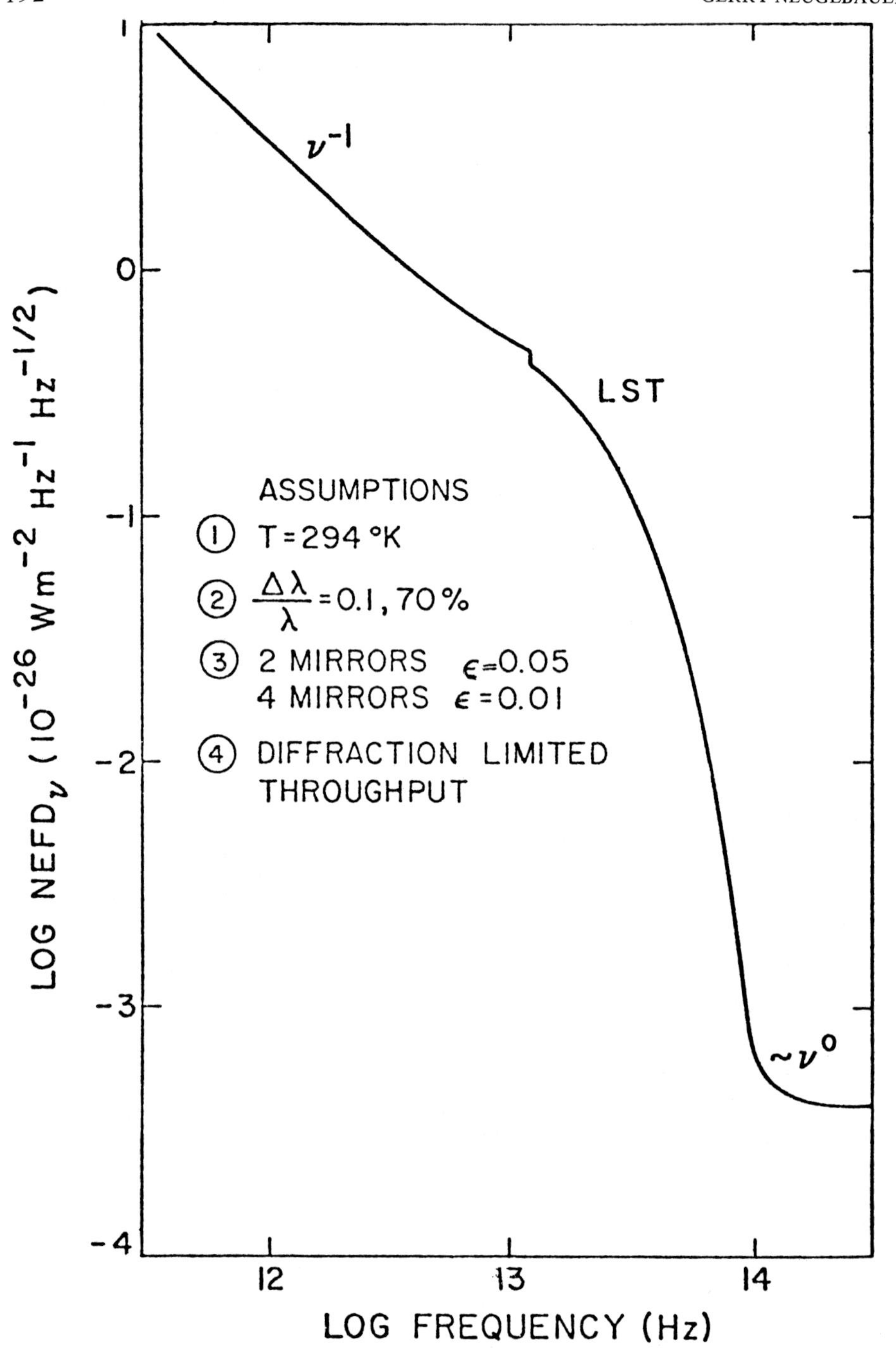

Figure 4 - The data of Figure 3 are converted to noise equivalent flux densities.

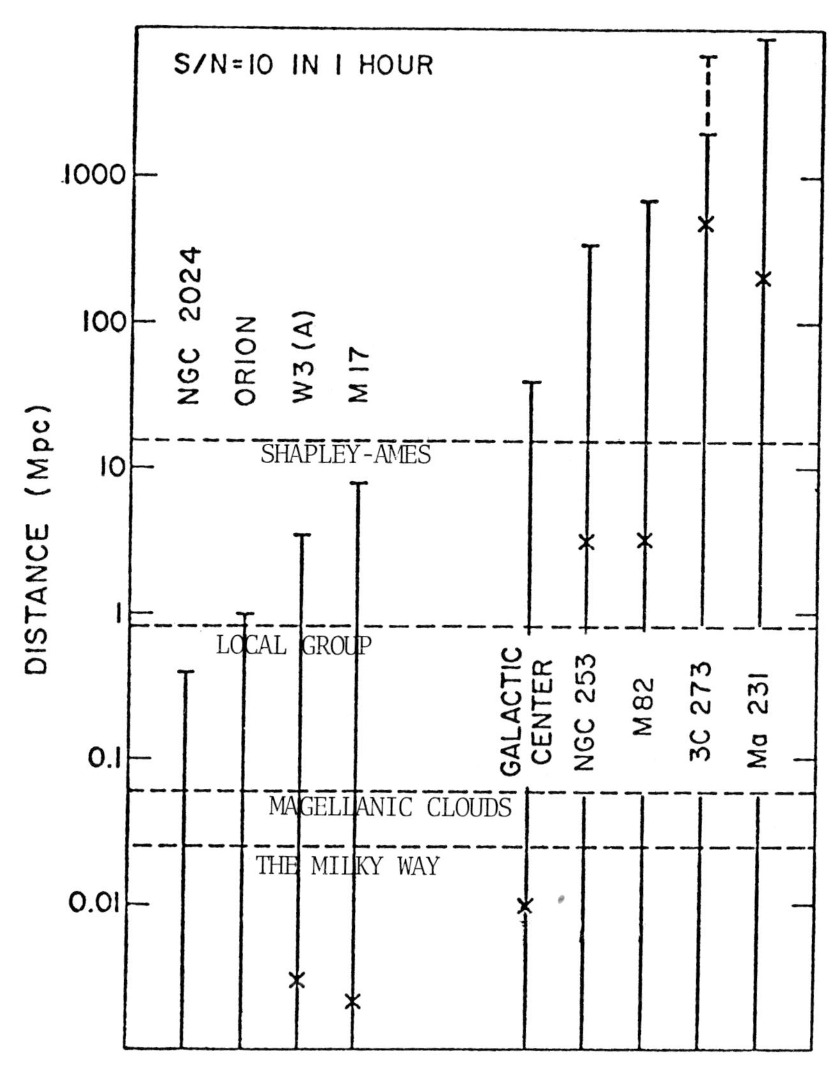

Figure 5 - The distance to which various objects can be seen at 100 μ by the (L)ST are shown.

per year in comparison with the full time available on the (L)ST. In fact, in all comparisons of the (L)ST with other space instruments, the large observing time available on the (L)ST cannot be ignored; it allows extensive projects to be carried out with thought and response based on results obtained during the mission.

A question of interest is, of course, sensitivity. In particular, how much is the sensitivity degraded by the warm telescope? Again the answer lies with the high angular stability since it is possible to reduce the flux on the detectors by decreasing the aperture to its diffraction limited size, thus decreasing the background fluctuations with respect to incoming signals. To illustrate the value of this, Figure 3 shows the noise equivalent power (NEP) set by the fluctuations in the mirror emission as a function of wavelength; a little reflection shows that for a diffraction limited aperture the throughput is proportional to the square of the wavelength independently of the telescope size of focal ratio. The point illustrated in Figure 3 is that the warm telescope does hurt around 10 μ, the peak of the Planck curve; longward of ~100 μ, however, the detector sensitivity is the limiting quantity and thus the emission from the warm telescope does not hurt the sensitivity. Another way of presenting these calculations is shown in Figure 4 which gives the predicted flux densities as a function of wavelength. The significance of Figure 4 can be seen in Figure 5 where the distance to which several known objects could be seen at 100 μ with the (L)ST is shown. The plot is due to D. Kleinmann; references for the data and extrapolations leading to the 100 μ fluxes are given in Hall et al. (1975).

Briefly, it is seen that an H II region like Orion could be studied if it were located in any of the galaxies making up the local group. In turn, all the galaxies in the Shapley-Ames catalog could be studied to see how ubiquitous or unique the infrared source in the center of the Galaxy really is. Energy distributions of quasars could be made with ease and 3C 273 the brightest quasi-stellar object could be studied to see whether variations reported in radio wavelengths extend to 100 μ.

In summary, at shorter wavelengths the main utility of the infrared photometer will be in the high spatial resolution of the (L)ST while at longer wavelengths the (L)ST will provide unequalled sensitivity as well. The infrared capability extends the wavelength coverage of the (L)ST by a factor of a thousand and truly belongs as one of the instruments in the spacecraft observatory.

Hall, R. T., Kelsall, T., Kleinmann, D. E. and Neugebauer, G. "Infrared Capabilities", The Space Telescope, NASA publication as presented at the 21st annual meeting of the AAS, Denver 1975.

A 1-METER CRYOGENIC TELESCOPE FOR THE SPACE SHUTTLE

F. C. Gillett

Kitt Peak National Observatory*
Tucson AZ 85726

1. INTRODUCTION

It has become clear in the last few years that infrared astronomy is ready for the leap into space. Indeed, the first steps in this direction have already been taken. NASA's airborne observatories, particularly the 91-cm telescope mounted in a C-141, and an expanded balloon program have opened up for study new and exciting wavelength regions inaccessible from the ground. Astronomical observations with cryogenically cooled telescopes carried into space by rockets have been made, confirming the very low infrared backgrounds outside the Earth's atmosphere [1] and the potential for high-sensitivity, high-speed observations [2].

There are two basic reasons for the appropriateness of a substantial space effort in infrared astronomy at this time:

(1) recent technological advances leading to long-lived cryogenic systems, plus development of photoconductive detectors with very high sensitivity under low background conditions primarily in the wavelength range to 30 μ; and

(2) infrared astronomy has matured to the point where a large number of very exciting scientific problems are under study, ranging from asteroid [3], satellite [4], and planetary [5] investigations to star formation and evolution [6] and the 3° background. These problems are among the most exciting in modern astronomy and all would benefit greatly from high-sensitivity observations above the Earth's atmosphere. Thus, a cooled

* Operated by the Association of Universities for Research in Astronomy, Inc., under contract with the National Science Foundation.

G. G. Fazio (ed.), Infrared and Submillimeter Astronomy, 195-206.

infrared telescope in space will surely provide a wealth of immensely valuable information over virtually the entire gamut of astronomical research.

The potential advantages for infrared observations from space are basically due to elimination of atmospheric absorption and emission, and a greatly reduced background photon flux which allows the use of much more sensitive detectors.

Cooling the telescope will leave emission from zodiacal dust particles as the dominant source of background radiation. Near 90° elongation in the ecliptic plane, the measured surface brightness of the zodiacal emission around 10 μ is about 6×10^{-11} w cm^{-2} $sterad^{-1}$ μ^{-1} with a color temperature of about 300 K [1]. This amounts to a reduction in background surface brightness by a factor 10^{6}-10^{7} from that of an ambient temperature telescope operating within or outside the Earth's atmosphere, and implies a potential improvement in detector sensitivity of $1/\sqrt{BG}$, or 10^{3} or more.

In order to investigate the present technical feasibility and develop a preliminary design of a cooled infrared telescope in space, NASA has recently initiated a preliminary design study for a 1-m cryogenically cooled infrared telescope for the space shuttle (SIRTF). This study [7] concluded that such a facility was indeed well within the range of present technology, and the following sections describe the preliminary design, its capabilities and limitations. Similar discussions of the preliminary design, from a slightly different point of view, have been presented previously [8,9].

2. PRELIMINARY DESIGN OF SIRTF

Figure 1 shows the radiative backgrounds in the shuttle environment. The zodiacal emission is taken from a model by Jarecke and Wyett [10]. Expected emission from 10^{12} molecules cm^{-2} of H_2O and CO_2 shuttle-associated contamination is also included [11], as well as lines corresponding to background-limited NEPs assuming 1-m clear aperture, 1 arcmin square field of view, and a 10 μ bandpass and the emission from 10 K and 20 K blackbodies. Notice that the required temperature of the telescope depends primarily upon the performance of the long-wavelength detectors. The SIRTF specification was to allow background-limited detector performance at an NEP of 10^{-16} w $Hz^{-1/2}$ in any 10 μ band between 30-200 μ looking into a 1 arcmin field of view. To satisfy this requirement, the telescope temperature must be $\lesssim$18 K. However, at this temperature the thermal emission of the telescope is much greater than the zodiacal or H_2O emission for wavelengths greater than 40 μ. If more sensitive long-wavelength detectors become

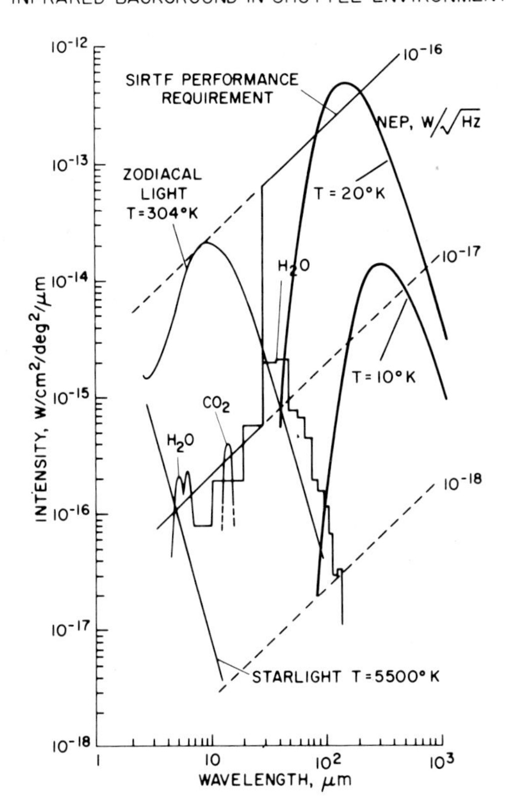

FIG. 1. Radiation backgrounds in the shuttle environment [9].

available in the near future (as seems likely), it may be desirable to decrease the telescope temperature to allow improved detector performance. Such flexibility has been built into the SIRTF design.

In addition to being cold the telescope must have very good out-of-field rejection, since both the Sun and Earth are extremely bright sources and the telescope will be operating while pointed as close as 30° from the Earth's limb and 45° from the Sun. Of course the telescope can never look directly at the Sun or Earth because the consequent large aperture loading would rapidly heat the telescope and consume the cryogen.

Spatial chopping is required to cancel the residual background while observing dim sources and eliminate the effect of low-frequency drifts inherent in many infrared detectors. It is important to note that, although the spatial chopping included in SIRTF is similar to that commonly done on ambient temperature telescopes, the situation is vastly different. For an ambient temperature telescope the background flux is invariably much larger than the source flux, while in SIRTF the background flux into a diffraction-limited beam at 10 μ, for example, is the same as that from a source of magnitude [N] = +9.5; i.e., about a factor of 10 fainter than any source yet detected at this wavelength.

Another important SIRTF concept has to do with instrumentation. Specific instruments have not yet been decided upon, and in order to retain flexibility in this area, where technology is advancing very rapidly, the SIRTF telescope has been designed to include a multiple instrument chamber (MIC) which can accommodate up to six instruments in the focal plane on each flight with the possibility of changing instruments between flights.

Figure 2 shows a schematic of the proposed optical system, a double-folded Gregorian configuration with a 1.16-m diameter primary mirror, diffraction limited at 5 μ (i.e., geometrical image diameter <2 arcsec) over a 15 arcmin clear field of view.

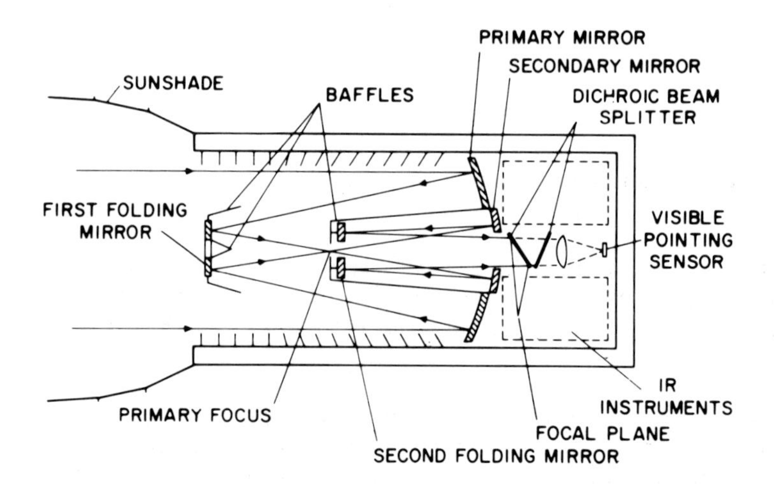

FIG. 2. Schematic of SIRTF optical system [9].

In this system light is first reflected off the primary mirror; then off the first folding mirror, forming a real image of the sky at the primary focus; then off the secondary mirror, forming a real image of the primary mirror at the second folding mirror; and finally, after reflection off a dichroic mirror (visible transmitting, infrared reflecting), forming a second real image of the sky in the multiple instrument chamber.

Out-of-field rejection is accomplished by careful baffling of the telescope, including a retractable parabolic sunshade, baffled telescope walls and mirrors, and stops at both the primary focus and image of the primary mirror.

The second folding mirror is convenient for spatial chopping and fine guiding. Tilting this mirror does not move the beam on the primary mirror, thus eliminating scan noise due to off-axis radiation scattered from condensed material or dust on the primary. Also, no defocusing or coma is introduced by tilting the second folding mirror, so the image remains diffraction limited even with large chopper throws (up to 15 arcmin).

The telescope is cooled by supercritical helium gas (initially at about 5 K and 5 atm pressure) stored in external tanks. Figure 3 shows a cutaway view of the telescope surrounded by three radiation shields and supported on conical fiberglass standoffs. The cold He gas is circulated through tubes cooling first the telescope walls, optics, and instrument chamber, then the radiation shields, and finally is vented into space through the telescope as a purge to impede the flow of contaminants onto the cold surfaces. At low flow rates the telescope temperature is maintained at about 17 K and 1 to 2 160-kg tanks are

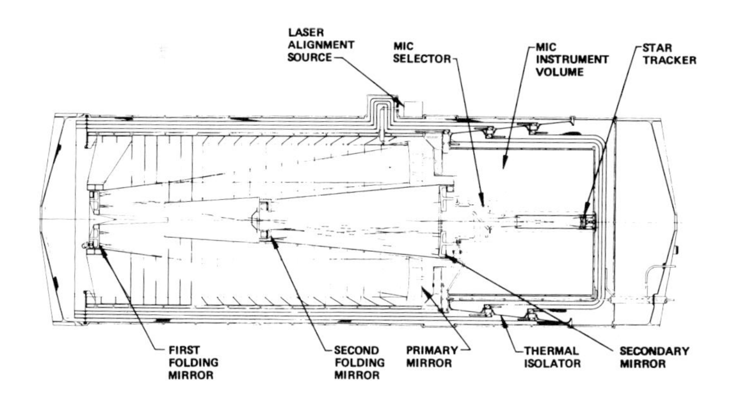

FIG. 3. Cutaway view of SIRTF telescope [7].

sufficient for a 30-day mission; at high flow rates the telescope temperature is maintained at about 12 K and 4 to 5 160-kg tanks would be required for a 30-day mission.

The optics are made of beryllium because of its attractive physical properties; specifically, good dimensional stability with time and after thermal cycling, and a high stiffness-to-weight ratio. The present capability for producing Be blanks limits the size of a finished primary mirror to about 1.6 m, which is the current upper limit for telescopes of this type.

Infrared photons are directed to one of up to six instruments by means of a rotatable dichroic mirror, while visual photons are transmitted to a CCD array used for acquisition and fine guidance. Instruments will generally be operated at temperatures in the range 12 K to less than 1 K. The multiple instrument chamber is designed to provide different degrees of cooling for each instrument, and some instruments may contain their own cryogenic supply for achieving very low temperatures.

Figure 4 shows a sketch of the telescope as it might appear on a spacelab mission. The telescope is shown mounted on the European Space Agency instrument pointing system (IPS), although the preliminary design study included a dedicated mounting as well. The IPS is a three-axis gimballed system, equipped with star trackers, providing pointing to 1 arcsec rms. Also shown are two (one hidden) super critical He tanks attached to the telescope body and a cryogenically cooled telescope cover, which provides a vacuum-tight seal for the front of the telescope during prelaunch, launch, and reentry phases of the mission.

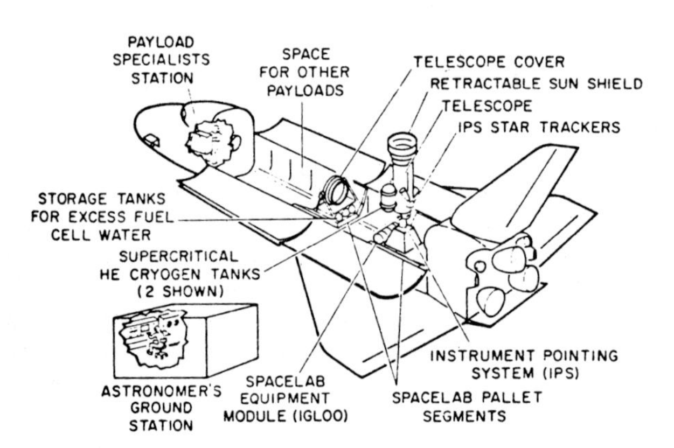

FIG. 4. SIRTF inflight configuration on the shuttle [9].

3. CONTAMINATION AND OPERATIONS

SIRTF operations are being planned to provide maximum observing time of astronomical objects, consistent with the necessary Earth and Sun avoidance and with minimum interference from contamination.

The most important shuttle-related contaminants are water vapor and dust particles. The emission from 10^{12} molecules cm^{-2} of water vapor is shown in Figure 1 and is considered to be the upper limit to the allowable column density of water [13]. Ambient temperature (200-300 K in near-Earth orbit) dust particles radiate strongly in the infrared. Particles as small as 5 μ in diameter can be detected as far away as a few kilometers; thus, transit of the line of sight by dust particles this large or larger must be minimized, preferably down a few per orbit [11].

There are a number of sources of shuttle-related contamination (solid particles and infrared active molecules); for example, outgassing from shuttle surfaces, exhaust and unburned fuel from attitude-control thrusters, excess water from fuel cells, and molecules and particles from vents. Every effort is being made to reduce and/or control this contamination, including choice of surface materials, use of clean-room techniques during prelaunch phases, manual control of vents, on-board storage of fuel-cell water and wastes, and selection of shuttle orientation requiring minimal use of thrusters during observations. Since the importance (relative and absolute) of the various sources of shuttle-related contamination is very difficult to calculate in advance, a contamination monitor is being planned for the earliest shuttle flights in order to evaluate the contamination due to water vapor and particles, and the various reduction and control schemes.

Figure 5 shows two possible orbiter orientations. In either case the telescope must acquire a new object 3 to 4 times per orbit because the orbital motion of the shuttle brings a given object too close to Earth (within 30° of the limb) in about 1/4 of an orbit. Slewing motions of the telescope will be accomplished using the telescope mount rather than reorientation of the shuttle itself. Figure 5a shows the orbiter orientation which gives maximum sky coverage; however, this orientation is unstable and frequent thruster firing may be required in order to maintain this orientation. A stable orientation is shown in Figure 5b. About 42% of the sky is observable with this orientation, which can be maintained with very infrequent use of the thrusters.

(a)

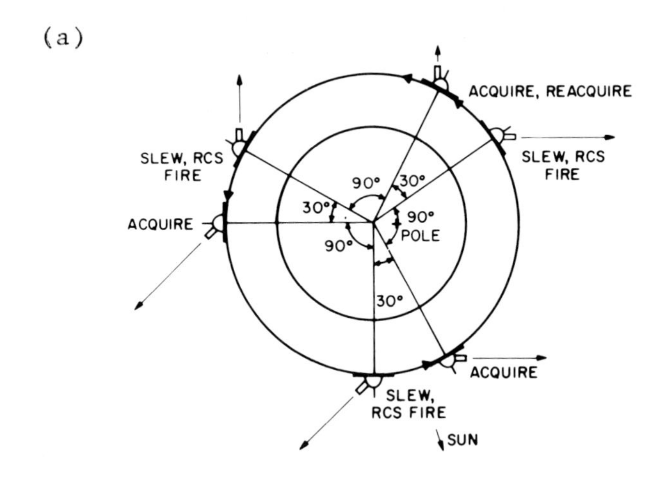

(b)

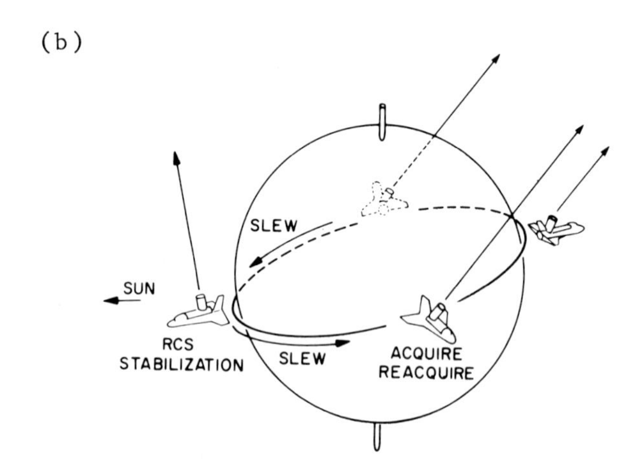

FIG. 5. Orbital orientation. (a) maximum sky coverage; (b) maximum viewing time between thruster firings.

4. SCIENTIFIC USE OF SIRTF

The possible scientific uses of SIRTF have been discussed by a number of groups [12,13,14], and cover a very broad range of astronomical problems.

SIRTF is particularly well suited for photometry of dim and/ or extended sources over a wide range of wavelengths. As pointed out earlier and presented in Table 1, the telescope and sky background for SIRTF are so small that virtually every presently known infrared source is bright compared to this background. Thus, total flux measurements rather than differential spatial measurements can be made on these and many other presently

TABLE 1

SIRTF Performance

	Wavelength	10 μ		30 μ		100 μ	
	Beam Diameter	1'	[a]4.3"	1'	[a]13"	1'	[a]43"
Zodiacal Emission[b]	Photons $cm^{-2}\,sec^{-1}\,\mu^{-1}$	200	1.0	70	3.3	3.7	1.9
	Jy (10^{-26} w $m^{-2}\,Hz^{-1}$)	1.4	0.007	1.4	0.07	0.25	0.12
	Magnitude	+3.6	+9.5	+1.2	+4.6	+0.2	+1.0
	NEP (10^{-17} w $Hz^{-\frac{1}{2}}$)[c]	30	2	6	1.4	10	10
Flux Limits[d]	$\Delta\lambda = 10\mu$: mJy (10^{-29} w $cm^{-2}\,Hz^{-1}$)	1.0	0.07	2	0.4	30	30
	$\Delta\lambda = 10\mu$: Magnitude	+11.5	+14.5	+8.5	+10	+5	+5
	$\lambda/d\lambda = R^{e}$, $n = \Delta\lambda/d\lambda$	4×10^{-7} Rn(Jy)				$\sim 7 \times 10^{-3}$ R(Jy)	
Integrating Time[f]	W3-IRS 5, $R = 10^{4}$	36 sec[g]		24 sec		-	-
	$F_{\nu} = 0.1$ Jy, $R = 100$		24 sec		7 sec	500 min	
Max. Distance (MPC)[h]	Galactic Center	30	120	20	40	100	100
	α Orion	0.7	2.7	0.2	0.2	-	-
	W51-IRS 2	3	12	6	12	14	14

[a]Diffraction limited beam diameter 2.5 λ/D.

[b]300 K grey body, emissivity = 6×10^{-8}.

[c]10,30 μ - background noise limited $\eta = 0.3$, $\Delta\lambda = 10\ \mu$.
100 μ - NEP = 10^{-16} w $Hz^{-\frac{1}{2}}$, $\Delta\lambda = 10\ \mu$.

[d]S/N = 10, 10-min. integrating time, transmission = 0.5.

[e]10,30 μ - source photon noise limited $\eta = 0.3$, transmission = 0.1.
100 μ - NEP = 10^{-16} w $Hz^{-\frac{1}{2}}$, trans. = 0.1, full multiplexing advantage.

[f]S/N = 10, spectral range ($\Delta\lambda$) = 10 μ.

[g]Outside absorption feature.

[h]S/N = 10, $\Delta\lambda$ = 10 μ, 10-min. integrating time.

unknown sources. This capability is unique to a cryogenically cooled infrared telescope in space and will be invaluable in the study of extended sources such as external galaxies and dark-cloud/molecular-cloud complexes.

The broadband photometric capabilities of SIRTF will not be discussed beyond saying that SIRTF will be, by a very substantial margin, the most powerful infrared telescope available now or envisioned for the near future. Limiting broadband fluxes are given in Table 1 together with the distances at which sources as bright as the galactic center (some galactic nuclei are 1000 times more luminous), α Ori (a typical M supergiant), and W51-IRS 2 (a typical infrared source associated with a compact H II region and a molecular cloud) could be seen. Clearly, broadband photometry using SIRTF will provide invaluable information on a wide variety of scientific problems involving both extended and faint sources, including energetics of extragalactic sources, and star formation and evolution within our own Galaxy as well as other galaxies.

Scientifically, high resolution spectroscopy using SIRTF will be very important because, in addition to the sensitivity of instruments on SIRTF, most of the vibrational and rotational transitions of molecules, including those of H_2, occur in the infrared. At present molecular astronomy is carried out almost entirely at millimeter wavelengths. However, for many purposes the infrared is better suited for this work because (1) the strongest molecular transitions occur in the infrared; (2) there exist large numbers of reasonably bright infrared sources within molecular clouds, which allows one to do absorption-line spectroscopy as well as emission-line spectroscopy; and (3) even with relatively small instruments, much higher spatial resolution can be achieved in the infrared because of the much shorter wavelengths. The ability to do absorption spectroscopy will be critically important, since it allows the detection of molecules in very cool shells or clouds; i.e., clouds where there is no appreciable excitation of the molecules. The basic requirement of interstellar or circumstellar molecular astronomy, particularly for absorption-line spectroscopy, is high spectral resolution; i.e., the bandwidth of a spectral element must be the same order as the line width, which requires $R = \lambda/\Delta\lambda = 10^4\text{-}10^5$. Table 1 also shows the expected capabilities of SIRTF for spectroscopy. For sources bright enough to do high resolution spectroscopy, the flux from the source dominates the background, so that, for a photon noise-limited detector, the time required to obtain a spectrum is inversely proportional to the flux from the object rather than to the square of the flux from the object. This also means that there is no great multiplexing advantage when observing bright sources. However, Michelson interferometers will still have a large throughput advantage, and multiplexing will be important for weak sources and in spectral ranges, or at spectral resolutions, where detectors are not photon noise limited. As seen in Table 1, SIRTF will have the capability to obtain very high resolution spectra of many of the brighter sources in molecular clouds over a very wide range of wavelengths (W3-IRS 5 being a typical example of this type of source), and many weaker sources at the same resolution, but restricting the wavelength coverage to individual features of interest. This means that the capability will be available to study the molecular composition and abundances, isotopic ratios, temperature, and dynamics of a large number of regions where it is thought that star formation is taking place. This information is vital to our understanding of the earliest stages of stellar evolution and may even hold the key to the origin of life in our Galaxy.

The ability to measure Doppler shifts of emission or absorption features will also be very important in that such information can be used to estimate distances and thus luminosities, which, for many galactic infrared sources, are very difficult to estimate in any other way.

5. SUMMARY

This paper has attempted to sketch briefly the following picture: (1) there is an enormous potential for doing infrared astronomy from space; (2) we are in a position to exploit this potential both technologically and scientifically; (3) SIRTF is an extremely powerful and flexible observing tool, which has the capability (in some cases nearly unique) to make critical observations in many of the most exciting areas of current astronomical research.

ACKNOWLEDGEMENTS

The author would like to thank F. Witteborn and L. Young of NASA, Ames, for numerous discussions and permission to use some of their material.

REFERENCES

[1] Soifer, B. T., Houch, J. R., and Harwit, M.: 1971, Astrophys. J. Letters 168, L73.
[2] Walker, R. G. and Price, S. D.: 1975, AFCRL TR-75-0373.
[3] Morrison, D.: 1974, Astrophys. J. 194, 203.
[4] Gillett, F. C.: 1975, Astrophys. J. Letters 201, L41.
[5] Gillett, F. C. and Forrest, W. J.: 1974, Astrophys. J. Letters 187, L37.

[6] Strom, S. E., Strom, K. M., and Grasdalen, G. L.: 1975, Ann. Rev. Astr. Astrophys. 13, 187.
[7] Shuttle Infrared Telescope Facility (SIRTF) Preliminary Design Study Final Report, June 1976, Hughes Aircraft Co.
[8] McCarthy, S. G., Young, L. S., and Witteborn, F. C.: 1975, Amer. Astronautical Soc. No. 75-284.
[9] Witteborn, F. C. and Young, L. S.: 1976, Amer. Inst. of Aeronautics and Astronautics No. 76-174.
[10] Jarecke, J. and Wyett, L. M.: 1975, Hughes Aircraft Co. Interdepartmental Correspondence No. 27735/180.
[11] Simpson, J. P. and Witteborn, F. C.: 1976, in preparation.
[12] Final Report of the Payload Planning Working Groups, NASA, Goddard Spaceflight Center, Greenbelt, MD, May 1973, Vol. I.
[13] Scientific Uses of the Space Shuttle: 1974, National Academy of Sciences, Washington, DC.
[14] Space Science Board Study on Infrared and Submillimeter Astronomy: 1975, National Academy of Sciences, Washington, DC.

LIRTS: A LARGE INFRARED TELESCOPE FOR SPACELAB

A.F.M. Moorwood

European Space Agency
Astronomy Division, Space Science Department
European Space Research and Technology Centre
Noordwijk, The Netherlands

SUMMARY

ESA is currently conducting the Phase A study of a large infrared telescope for Spacelab. The preliminary design of a 2.8 m aperture, modified Ritchey-Chrétien telescope can be accommodated in the short module or pallet only Spacelab configurations. Pointing is by means of the Instrument Pointing System. The facility has outstanding potential for a wide range of astronomical observations in the far infrared.

INTRODUCTION

The European Space Agency is currently studying the feasibility of an infrared telescope of about 3 m diameter for Spacelab. Provision of such a facility would offer enormous advantages over existing aircraft and balloon-borne systems for detailed studies of astronomical sources in the far infrared.

LIRTS would have wide application for studies of cool objects including planetary atmospheres, early and late type stars, the interstellar medium and external galaxies. Recent and unexpected discoveries have shown that observations in this spectral region are vital to our understanding of many important phenomena, including that of star formation, which occur in visually obscured regions.

The concept of a large ambient temperature telescope which fully utilizes the mass and volume capability of Spacelab was recommended for further study in 1974 (1) by the Infrared

G. G. Fazio (ed.), Infrared and Submillimeter Astronomy, 207-223.

Mission Definition Group set up by ESA. Several options were considered by this group which was supported by a preliminary industrial study performed by Engins Matra, France (2). The primary scientific objectives foreseen are those in the wavelength range from about 30μm to 1 mm which will most benefit from the freedom from atmospheric limitations and the high spatial and spectral resolution capabilities of LIRTS. For these observations, requiring diffraction limited fields of view, the gain to be expected from cryogenic cooling of the telescope optics is considered of less importance than the gain in spatial resolution achieved by use of the largest possible diameter. The objectives are complementary however to those of NASA's proposed SIRTF telescope (3) which is much smaller in diameter but which uses cryogenically cooled optics to permit full use of the much lower N.E.P. detectors available particularly below 30μm.

Specifications for the Phase A study of LIRTS were drawn up around the middle of 1975 and the industrial study was awarded to Engins Matra, France (with sub-contractors TNO-TPD, Netherlands and Reosc, France) at the end of 1975 following a competitive tender action. Scientific guidance of the study is the responsibility of a science team composed of ESA staff and consultants which has also generated the typical requirements anticipated for the focal plane instruments.

Evaluation of the technical results of the study is being made with the help of a team of ESA engineers.

This paper summarizes briefly the results of the study so far with particular emphasis on the scientific objectives and the performance characteristics of the preliminary LIRTS design.

SCIENTIFIC OBJECTIVES

The earth's lower atmosphere is strongly absorbing between 20μm and 1000μm and astronomical observations are impossible from the ground throughout most of this wavelength range. Exploratory measurements with aircraft and balloon-borne systems have already revealed the importance of this spectral region however for the study of a wide range of objects.

Cold clouds of molecules and dust in our Galaxy radiate predominantly in the far infrared because their temperatures are typically below 100 K. These regions provide the sites for star formation which can only be observed at infrared wavelengths. The formation of the molecules themselves also appears to be intimately connected with the presence of the dust which is producing the bulk of the infrared emission.

The predominant energy output of HII regions has also been found to be in the far infrared. Again it is the radiation from solid particles which is measured but the primary energy sources here are the recently evolved O and B stars which are at temperatures above 25,000 K but whose ultraviolet and visible output is absorbed by dust. In most cases the absorption is sufficient to completely obscure both the stars and the nebular emissions at visible wavelengths. Infrared observations of the interstellar medium are vitally important therefore to improve our understanding of the fundamental process of star formation and evolution and many related phenomena occurring in the nebulae in which they form. The centre of our Galaxy presents a similar challenge. This region is a prominent and extended complex of infrared sources but is completely obscured at visible wavelengths by the concentration of interstellar dust in the galactic plane.

Whole galaxies have also been found to radiate the bulk of their energy in the far infrared and many more, which have so far only been detected in the near infrared from the ground, exhibit spectra which are still rising at 20μm. These include active (Seyfert, Markarian etc.) and peculiar as well as nearby normal galaxies. The origin of the enormous infrared luminosities is difficult to explain and the infrared may well be a key region for increasing our understanding of the evolution of galaxies.

In addition to radiation from dust the infrared region is rich in atomic and molecular transitions which will give unique information concerning the chemical composition, physical conditions and dynamics in a wide variety of situations. These include the objects mentioned above and also planetary atmospheres where molecular line measurements can be used additionally to determine the thermal profiles. In visually obscured HII regions and also HI regions the atomic and ionic abundances can be determined through measurements of ground state fine structure lines. The far infrared is a particularly important region for molecular studies because it contains the vibration-rotational and pure rotational transitions of molecules including many which are non-polar and possess no radio transitions. An important example is H_2 which can only be detected in dense molecular clouds in the infrared. Very high resolution spectroscopy will probe the physical conditions, excitation processes and dynamics of the clouds and will give unique information on the molecular formation mechanisms and the chemistry of the sources. These observations will greatly extend the studies currently possible using groundbased radio and mm wave systems.

Existing airborne instruments are limited by the presence of residual atmosphere which both emits and absorbs selectively

TABLE 1 LIRTS SCIENTIFIC OBJECTIVES

OBJECT	ASTROPHYSICAL INTEREST	OBSERVATIONS
Planetary atmospheres	Cloud structure. Thermal profiles. Chemical composition.	λλ 10 μm - 300 μm. Spatial resolution. Molecular spectroscopy ($\lambda/\Delta\lambda > 10^3$).
Stars and 'starlike'	Circumstellar envelopes. Mass loss. Composition. Novae shells.	λλ > 10 μm Spectrophotometry and polarimetry. Time variability. Spectroscopy ($\lambda/\Delta\lambda > 10^3$).
Molecular clouds. HII and HI regions. Galactic centre.	Nature and distribution of dust. Protostars. Stellar evolution. Development of HII regions.	Multiband photometry ($\lambda/\Delta\lambda \sim 1\text{-}20$). λλ 20 μm - 1 mm. High spatial resolution mapping. Polarimetry.
	Chemical composition. Physical conditions. Molecular formation. Cloud dynamics. Masers. Cooling mechanisms.	High ($\lambda/\Delta\lambda > 10^3$) and very high ($\lambda/\Delta\lambda > 10^5$) atomic and molecular spectroscopy. λλ 20 μm - 1 mm.
Galaxies	Emission mechanisms. Structure. Physical conditions. Evolution.	Multiband photometry λλ 20 μm - 1 mm. High spatial resolution. Spectroscopy ($\lambda/\Delta\lambda > 10^3$). Time variability.

in molecular lines and whose transparency fluctuates. The telescopes flown are also relatively small (< 1m) and consequently have inferior spatial resolution compared with telescopes available on the ground in other wavelength regions. The large size of LIRTS offers a substantial improvement in this respect and the primary objectives foreseen are those where good spatial resolution is important for the interpretation of both photometric and spectroscopic data. A number of specific objectives and the types of measurement required are listed in Table 1. The list is not intended to be comprehensive nor is it possible to state categorically where the priorities will lie in the 1980's. It is indicative however of the wide range of astrophysical studies for which LIRTS could be used.

FOCAL PLANE INSTRUMENTATION

An important aim of the LIRTS design is to retain maximum flexibility for the accommodation of a wide range of focal plane instruments. To establish a representative set of requirements for the Phase A telescope design the science team has considered three model instruments compatible with the scientific objectives outlined above. These are a multiband photometer/polarimeter, a Michelson interferometer ($\lambda/\Delta\lambda \sim 10^4$) and a submillimetre heterodyne line receiver ($\lambda/\Delta\lambda > 10^5$).

A common requirement of all three instruments is for cyrogenic cooling of the detectors and some optical components.

In the case of the photometer the complete instrument is cooled below 2K by liquid helium. The heat loads and cooled volumes have been found to be similar enough in the three cases however to consider development of a single dewar design compatible with these and presumably other instruments. Preliminary sizing of the dewar is compatible with at least two of the model instruments being flown on any one mission. The masses, thermal control, data handling and control of these instruments present no major problems.

TELESCOPE

Design

Phase I (system assessment) and Phase II (system design) of the industrial study have been completed at the time of writing (May 1976) and the final phase devoted to producing a cost and development plan is now in progress. The assessment phase was concerned primarily with a number of performance and Spacelab accommodation trade offs which were performed prior to establishing a baseline for a more detailed system design. The essential features and performance characteristics of the 2.8 m aperture telescope selected for more detailed study have now been established.

Figure 1 gives an overall system view. The telescope has a Cassegrain configuration and the focal plane instruments are located in the volume of about 7 m^3 provided behind the primary mirror. The end of the instrument chamber mates with the mounting flange of the Spacelab Instrument Pointing System (IPS) gimbal assembly. One boresighted star tracker is mounted externally on the telescope structure and close to both the primary mirror support ring and the clamping plane. It is optically baffled along the length of the telescope. Two roll sensors are located on the conical shell of the focal plane instrument chamber. The end of the telescope is sealed by a soft cover during ground handling and launch to prevent contamination of the optical surfaces and an extendable shield is provided to give improved on-orbit protection against stray light. The shield is canted to minimize the pointing direction constraints imposed by the Sun and the Earth.

The F/2 primary mirror has a diameter of 2.8 m and is made of ULE (Corning Glass). It is a lightweight construction comprising face and back plates fused to an 'eggcrate' core. The secondary mirror is slightly elliptical (0.335 x 0.36 m) and is formed by replication on a thin beryllium substrate mounted on stiffeners. This mirror can be held stationary or can be oscillated to provide space chopping. The dynamically balanced oscillating mechanism is coupled via a spring and damper arrangement to avoid torques on the telescope structure. An optical figure intermediate between Ritchey-Chrétien and tilted aplanatic has been chosen to give optimum image quality over an extended field when space chopping. The secondary mirror mount has five degrees of freedom allowing for centering, tilt and focus adjustments and a small collimating relescope located in the primary mirror central hole is used for sensing centering and tilt errors. An optical assembly in the focal plane instrument chamber is provided for focus error sensing and for alignment of the telescope and star tracker optical axes. The telescope beam is directed into this unit by a rotatable beam diverter mirror which is also used to switch between the infrared instruments. A beamsplitter is used to reflect part of the light from a reference star to the focus sensor while transmitting the remainder to a Vidicon camera and light intensifier which images a 10' field in the focal plane. A third optical arm can be used to feed light from an artificial star into the star tracker via the telescope and a corner reflector.

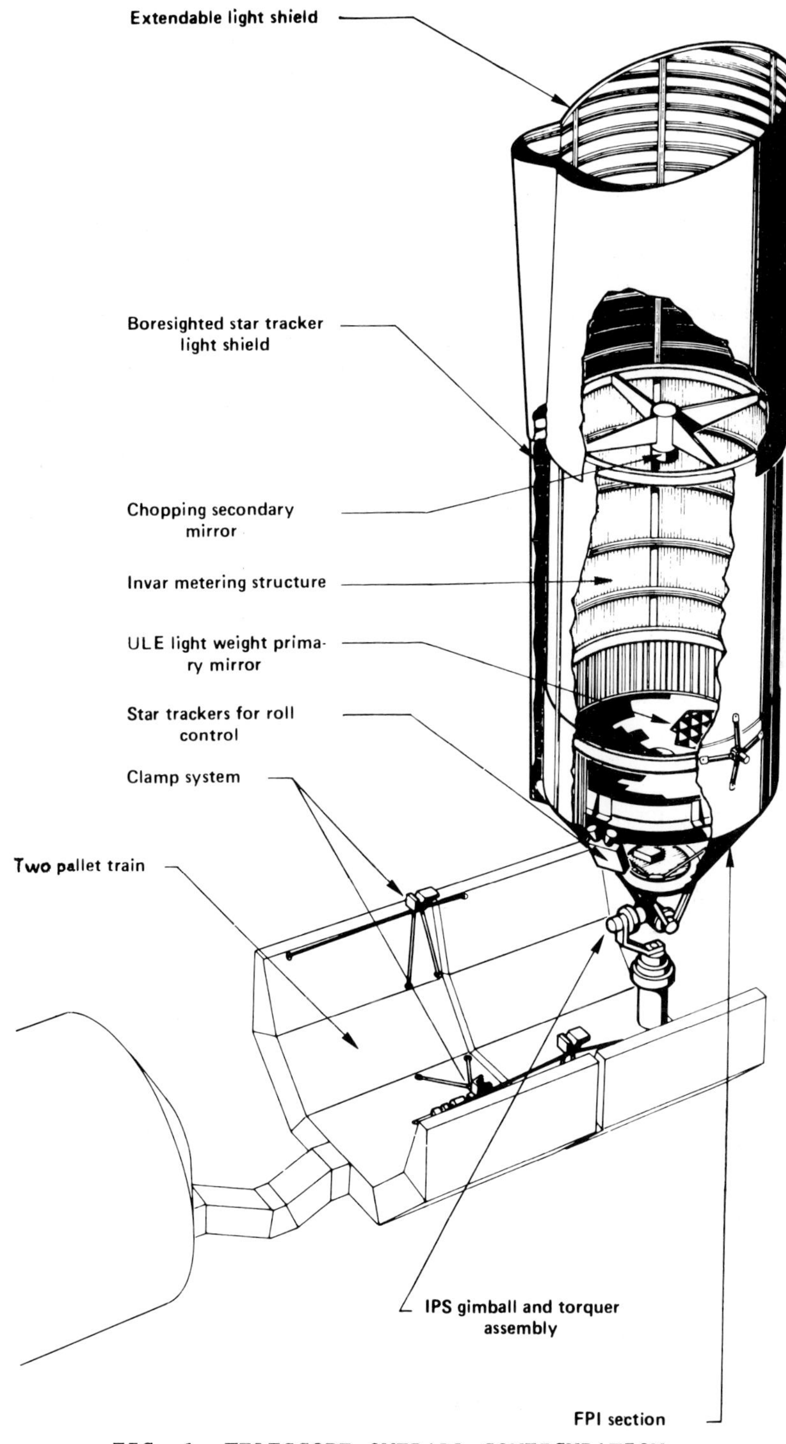

FIG. 1 TELESCOPE OVERALL CONFIGURATION

To avoid excess thermal radiation reaching the infrared detectors the secondary mirror is slightly undersized and the cross section of the supporting spider has been minimized. Conventional optical baffles have been avoided with the same purpose.

The mechanical metering structure between the primary and secondary mirrors consists of longitudinal invar members and corrugated skin panels and annular aluminum ring frames. The conical truss enclosing the focal plane instrument volume is of fibreglass and the extendable shield is formed of aluminum skin panels stiffened by aluminum rings which also provide baffling. The shield is locally extended to accommodate the boresighted star tracker.

The on-orbit thermal behaviour of the telescope is extremely important from the point of view both of its optical and noise performance and particular attention has been given in the design to minimizing temperature gradients across the primary mirror.

Fig. 2 shows the concept of the thermal design. A multilayer insulation blanket around the telescope minimizes temperature gradients across the structure and gradients across the primary mirror itself are further reduced by a metal isothermalizer plate which is radiatively coupled to the back of the mirror. The thermal conductivity of this plate is improved by four circular heat pipes. To avoid the thermal plate being affected by the primary ring it is conductively decoupled through an insulated support of glass epoxy and radiatively decoupled using multilayer insulation. The honeycomb plate used to support the focal plane instruments is also conductively and radiatively decoupled from the primary mirror assembly to minimize interference. Heat dissipation of the focal plane instruments is rejected by means of radiators on the conical shell. The only active thermal control element is a heater used to reduce the secondary mirror temperature fluctuations.

Performance

Performance figures derived under severe orbital assumptions have verified the capability of the above design to meet the scientific requirements. A summary of the main performance characteristics is given in Table 2.

At the shorter wavelengths some restriction below the maximum chopping throw and field of view is required to ensure diffraction limited performance. At 15 μm for example the system is diffraction limited over a field of 2' with a chopper throw of ± 2'. Over the far infrared range of prime interest however

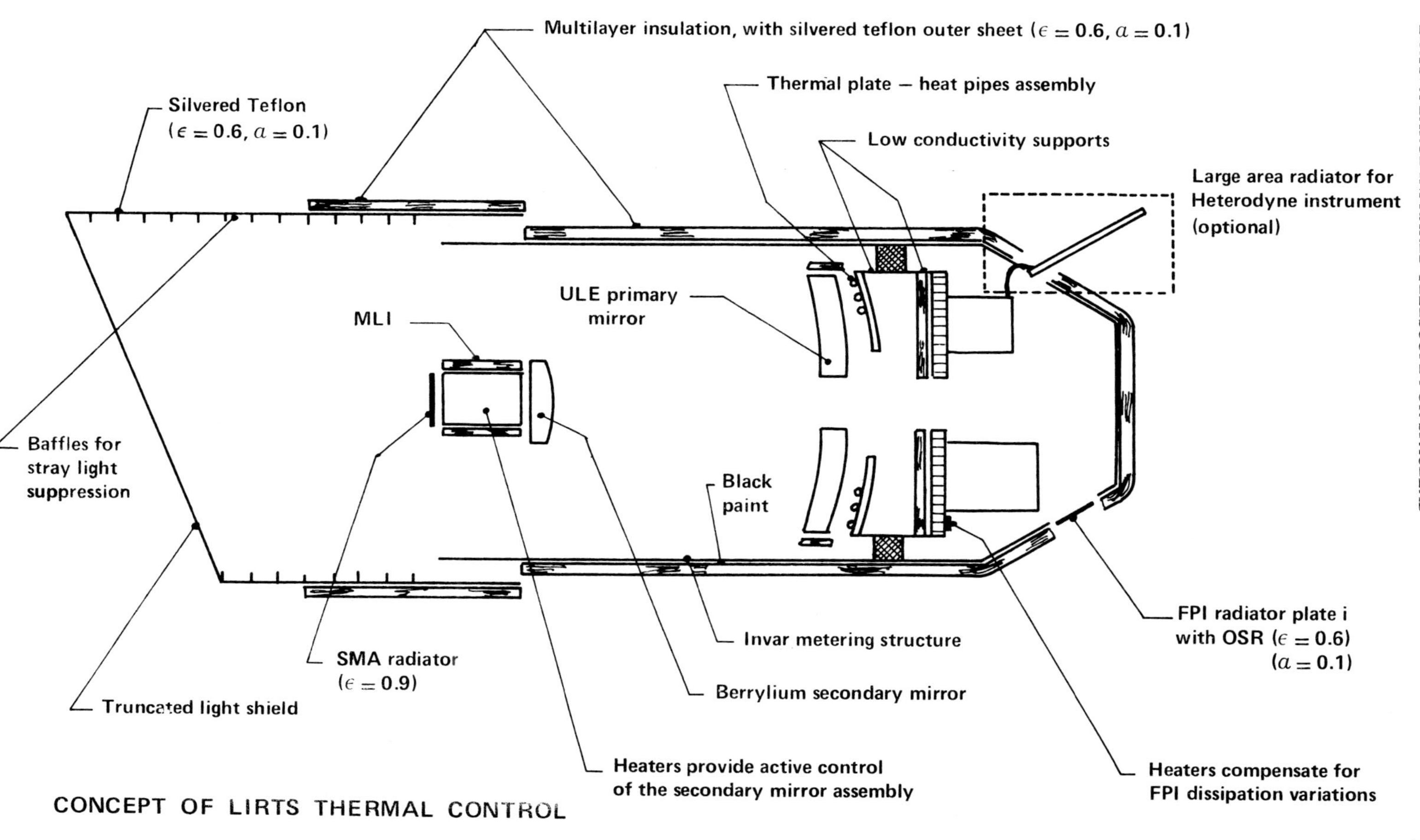

FIG. 2 THERMAL DESIGN CONCEPT

the system is diffraction limited over the maximum field of view and chopping throws specified. (The chopping frequency must be chosen below 80 Hz if the maximum throw of ± 10' is required.) The variations in imbalance signal during space chopping are dominated not by variations in the primary mirror gradient but by scattering of radiation emitted by the inside walls of the telescope. This radiation becomes unsymmetrical if a temperature gradient across the structure is produced by solar radiation entering the telescope. The overall sensitivity is unaffected however providing the optical axis is constrained not to point within about 45° of the sun.

The potential problems due to molecular and dust particle contamination which have been studied in relation to SIRTF are relatively unimportant for LIRTS. No increase in noise should be experienced providing the molecular column densities are below 10^{14} mols/cm^2. Molecular venting should be kept to a minimum but the only cold surfaces are inside the focal plane cavity and can be protected more easily than the mirror surfaces. Low orbits present some danger that the ambient gas pressure in the telescope could be too high for observations with the telescope pointing in the direction of flight. The visibility of dust particles ejected from the Orbiter/Spacelab is only a problem for particles larger than 100 μm which should be eliminated by the cleaning procedure.

The most serious potential noise source is due to charged particle hits on the detectors. Of most concern are trapped particles in the Van Allen belts and the South Atlantic Anomaly. Preliminary calculations show that the problem might be effectively solved however by a combination of screening and careful orbit selection.

SPACELAB ACCOMMODATION

LIRTS has been designed to be compatible with Spacelab in either the pallet only or short module configurations. Physical accommodation in the latter configuration is shown in Fig. 3. Two rigidly connected pallets are used, one supporting the telescope and its clamps and the other the IPS. This arrangement can also be reversed such that the IPS is located at the end of the cargo bay with the telescope extending in the direction of the module. Both configurations satisfy the orbiter center of gravity constraints but with a greater safety margin in the latter configuration. Table 3 gives the relevant physical parameters for LIRTS.

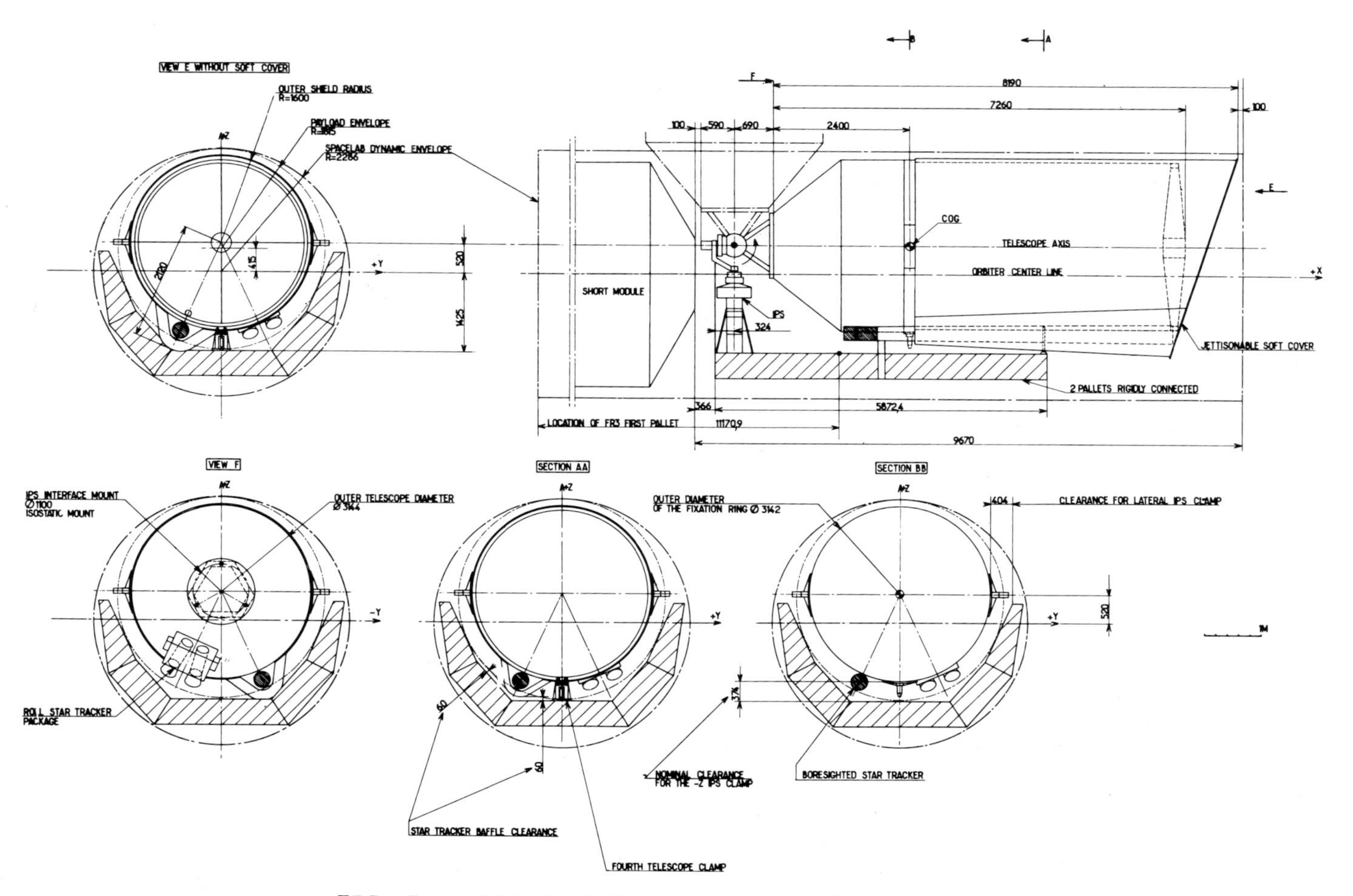

FIG. 3 ACCOMMODATION IN SHORT MODULE CONFIGURATION

TABLE 2 LIRTS PERFORMANCE CHARACTERISTICS

OPTICAL CONFIGURATION	MODIFIED RITCHEY CHRÉTIEN
Primary mirror diameter	2.8 m
Secondary mirror	0.335 x 0.36 m
Primary focal ratio	F/2
Overall focal ratio	F/15
Plate scale	5"/mm
Field of view	10 arc. min.
Diffraction limited wavelength	≃ 10 μm
Space chopping, throw	± 10 arc. min. (max.)
frequency	150 Hz (max.)
Sensitivity (N.E.P. for diffraction)	$< 3\cdot10^{-15}$ $W\cdot Hz^{-1/2}$ (10 μm - 100 μm)
Limited f.o.v. and $\Delta\lambda = 0.1\ \lambda$	$< 10^{-15}$ $W\cdot Hz^{-1/2}$ (> 100 μm)
Temperature fluctuation noise	$< 2\cdot10^{-17}$ W
Variation in imbalance signal over 30 mins. (max. chop)	$< 7\cdot10^{-17}$ W
Pointing (IPS), stability	1" (LOS) 3"(ROLL)
bias error	2"
raster field	≃ 2°
M_V, limit of focal plane camera	15

TABLE 3 LIRTS PHYSICAL PROPERTIES

Overall length stowed	9.47 m
Telescope outer diameter	3.14 m
Extendable shield length	5.66 m
LIRTS mass	3785 kg*
Payload mass	4510 kg+

*Includes 400 kg of focal plane instruments and 10% design margin.

+Includes IPS gimbal and clamp system.

On a pallet only mission LIRTS will represent about 48% of the total payload mass while its length is approximately 65% of that available on a 15 m pallet train. In the short module configuration LIRTS will represent approximately 75% of the available payload mass.

ORBITAL CONSIDERATIONS

The range of orbital altitudes (<460 km without OMS kit) and inclination 28.5° $\leqslant$ i < 57° achievable from the Kennedy Space Center (4) are adequate for LIRTS.

A low orbital altitude is favoured to reduce the charged particle effects while a high orbit would remove any problem of ambient gas condensation on cold surfaces in the focal plane. An orbit at around 350 km altitude is likely to be a satisfactory compromise.

On orbit there are constraints on the pointing direction with respect to the Sun and the Earth imposed by the need to maintain the level of thermal gradients and stray light within acceptable limits. The relevant parameters are the allowable elevation angles for the Sun and Earth with respect to the entrace plane of the shield. Beyond a certain value the unsymmetrical heating of the structure causes a degradation in the system noise performance as discussed earlier.

As the telescope can be orientated as required in roll the constraints on the directions which can be reached by the optical axis are greatly reduced by using a canted baffle. For a given mission the constraints generate two earth and one sun avoidance cap on the celestial sphere. Ideally the orbital elements should be chosen such that one of the earth avoidance caps, which are smaller, falls within the sun avoidance cap. Fig. 4 shows the avoidance caps for four seven day missions spaced by 3 months and at the same orbital inclination of 50°. The canting angle is 20° and the maximum elevations are 40° and 25° respectively for the earth and sun above the shield entrance plane. The boresighted star tracker baffling has been designed to allow use of stars down to 9th magnitude under these conditions. Maximum separation of the sun avoidance caps occurs for launch dates spaced by six months but the earth avoidance caps are then the same. Without changing the orbital inclination however it is still possible in principle to reach all points on the celestial sphere with two missions spaced three or four months apart. As it is anticipated that LIRTS would fly more than twice however it would be preferable to programme the observations in such a way that pointing to the limits set by the thermal requirements is unnecessary.

OPERATIONS

Integration of the focal plane instruments and loading of the cryogenic systems will take place prior to the installation of LIRTS in the Orbiter bay. The telescope will be continuously

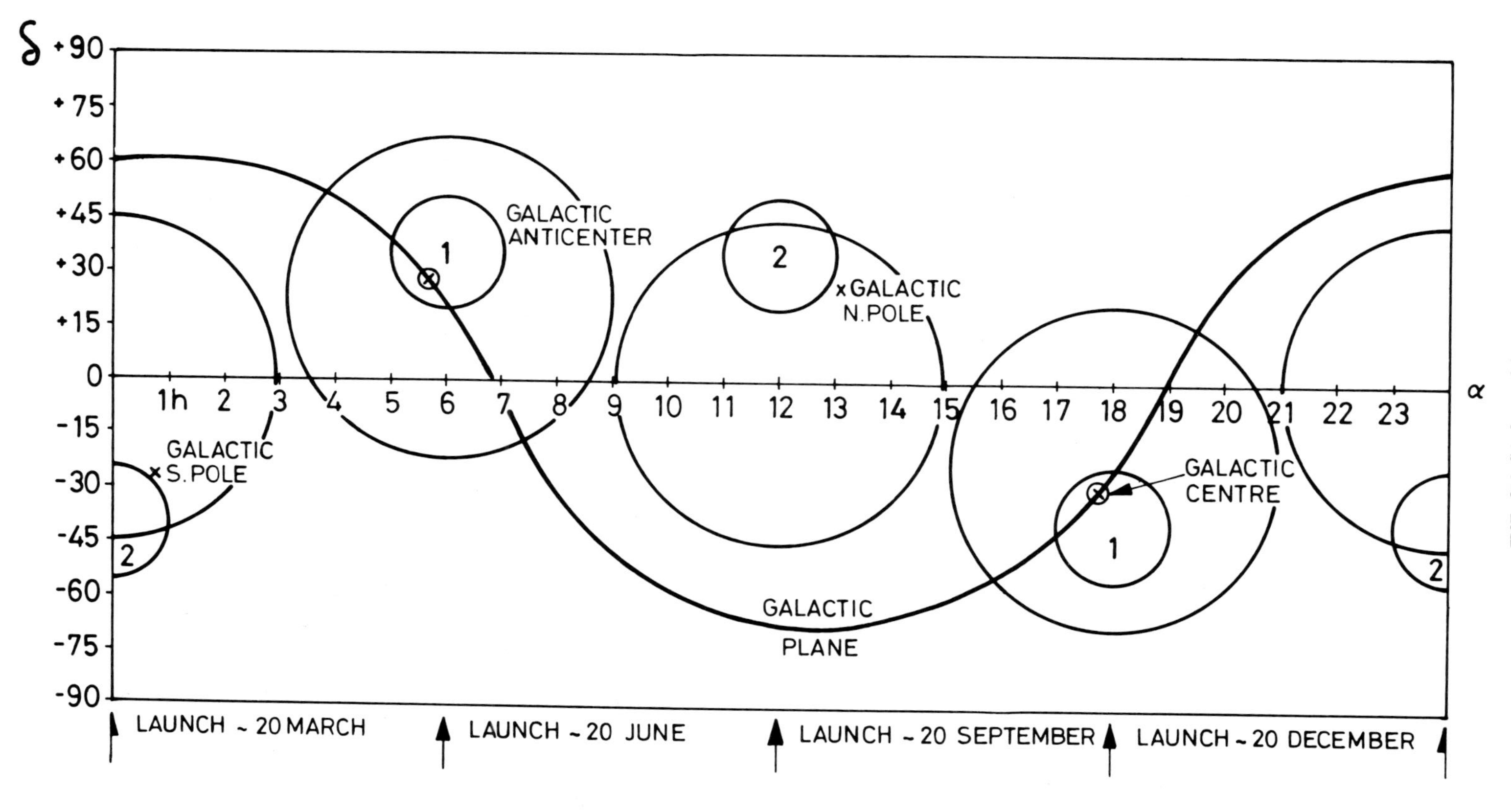

FIG. 4 SUN AND EARTH AVOIDANCE CAPS FOR FOUR SEVEN DAY MISSIONS

purged with dry gas until close to launch to protect the optical surfaces and topping up of the cryogens is desirable a few hours before launch. This may be a requirement for extended missions and a precaution in the case of nominal seven day missions.

Primary responsibility for the on-orbit operations is foreseen to rest with a trained payload scientist. After the cargo bay doors have been opened he or she will release the LIRTS clamps, lock the telescope to the IPS, remove the end cover, orientate the telescope to a reference star and extend the stray light shield. Centering, tilt and focus position of the secondary will then be adjusted, if necessary, and any bias between the telescope and star tracker axes will be measured. Preferably the reference star selected will also be bright enough in the infrared to allow any bias between the infrared and optical axes to be measured during the same sequence. After completion of these tasks the telescope will be orientated towards the first scientific target and the appropriate instrument selected. Re-pointing of the telescope will be required about four times/orbit. During the actual observations it is considered important that a scientific team on the ground has access to real or close to real time data and that they can communicate by voice link with the payload specialist. This team would be well qualified to assess the instrument performance and to provide a fast response in the case of unexpected results.

CONCLUSIONS

A large aperture telescope operating in space has enormous potential for a wide range of astronomical observations in the far infrared. The results to be expected would significantly increase our understanding of processes occurring in a range of cool objects as diverse as planets and distant galaxies. A preliminary phase design has been evolved for a 2.8 m diameter telescope which can be accommodated on Spacelab in its pallet only or short module configurations. Use is made of the Instrument Pointing System and the possibility for modifying or exchanging focal plane instruments between flights creates exceptional flexibility for the use of the facility. An important design requirement was to achieve diffraction limited performance at wavelengths longer than 30 μm with the largest possible telescope diameter. For the primary objectives foreseen, which utilize small fields of view and restricted spectral bandwidths, the fact that the telescope optics are at ambient temperature is not a major limitation on the achievable sensitivity. The modified Ritchey-Chrétian telescope selected meets the optical performance requirements with considerable margin and satisfactory noise performance can be achieved with a passive thermal design. Stray light constraints impose the requirement for at least two flights to reach all points on the celestial sphere. The system is foreseen to be operated on-orbit by a trained payload specialist in contact with a scientific team having access to real time data on the ground.

ACKNOWLEDGEMENTS

Technical data and Figures 1-3 have been taken from the Phase II report prepared by Engins Matra whose conscientious approach to this study has been greatly appreciated. Valuable technical support within ESTEC has been given throughout by Mr. R.J. Laurance and at various stages by other members of the Department of Development and Technology. Mr. H. Martinides advised on Spacelab payload aspects. I have benefited greatly in the Science Team from working with Dr. D. Lemke, Dr. J.P. Swings, Dr. G. Winnewisser and Professor R.E. Jennings who served as consultants, my colleagues in the Astronomy Division Dr. J.E. Beckman, Dr. T. de Graauw and Dr. P. Salinari and its chairman Dr. V. Manno.

REFERENCES

1. Large Infrared Telescope on Spacelab, 1974. Report on the Mission Definition Study. ESA, MAS(74)24.

2. A Large Infrared Telescope for Spacelab. Preliminary Analysis - Phase I report, 1975, MATRA R102/24/42/0 (ESRO Contract no. SC/52/HQ).

3. F.C. Witterborn and L.S. Young, 1976, AIAA 14th Aerospace Sciences Meeting, Paper no. 76-174.

4. Spacelab Payload Accommodation Handbook (preliminary issue May 1975).

ON THE DETECTABILITY OF MOLECULAR HYDROGEN WITH IRAS

K. W. Michel, T. Nishimura, H. Olthof

Max-Planck-Institut für Physik und Astrophysik
Institut für extraterrestrische Physik
Garching

ABSTRACT

It is proposed to observe the J = 2 to J = 0 transition of molecular hydrogen at 28 μ with a Fabry Perot ($\Delta\lambda \sim 100$ Å) in the survey mode as well as in the pointing mode of the Dutch satellite IRAS which is scheduled for launch in 1981. A discussion will be given on the astrophysical importance of this measurement as well as an outline of the technical feasibility.

G. G. Fazio (ed.), Infrared and Submillimeter Astronomy, 225.

INFRARED IMAGING SPECTROSCOPY

Richard B. Wattson

American Science and Engineering, Inc.
Cambridge, Mass., USA

ABSTRACT

The technology of CCD-type TV cameras combined with the technology of acoustically-tuned optical filters enables, for the first time, extremely sensitive and flexible infrared data acquisition combining both high spatial resolution and high spectral resolution. An instrument utilizing the above technologies is described and preliminary examples of data are shown in the form of spectral images of Saturn in the near infrared and a Minnaert analysis of Jupiter at these wavelengths. The imaging spectrometer which acquired the above data has a spatial resolution of 4", a spectral resolution of 15 Å and a spectral range of 0.65μ to 1.1μ. Recent improvements in camera and filter technology relevant to imaging spectroscopy are given to indicate what near and middle infrared data will be available to observers in the near future.

G. G. Fazio (ed.), Infrared and Submillimeter Astronomy, 226.

ASTROPHYSICS AND SPACE SCIENCE LIBRARY

Edited by

J. E. Blamont, R. L. F. Boyd, L. Goldberg, C. de Jager, Z. Kopal, G. H. Ludwig, R. Lüst, B. M. McCormac, H. E. Newell, L. I. Sedov, Z. Švestka, and W. de Graaff

1. C. de Jager (ed.), *The Solar Spectrum, Proceedings of the Symposium held at the University of Utrecht, 26–31 August, 1963.* 1965, XIV + 417 pp.
2. J. Ortner and H. Maseland (eds.), *Introduction to Solar Terrestrial Relations, Proceedings of the Summer School in Space Physics held in Alpbach, Austria, July 15–August 10, 1963 and Organized by the European Preparatory Commission for Space Research.* 1965, IX + 506 pp.
3. C. C. Chang and S. S. Huang (eds.), *Proceedings of the Plasma Space Science Symposium, held at the Catholic University of America, Washington, D.C., June 11–14, 1963.* 1965, IX + 377 pp.
4. Zdeněk Kopal, *An Introduction to the Study of the Moon.* 1966, XII + 464 pp.
5. B. M. McCormac (ed.), *Radiation Trapped in the Earth's Magnetic Field. Proceedings of the Advanced Study Institute, held at the Chr. Michelsen Institute, Bergen, Norway, August 16–September 3, 1965.* 1966, XII + 901 pp.
6. A. B. Underhill, *The Early Type Stars.* 1966, XII + 282 pp.
7. Jean Kovalevsky, *Introduction to Celestial Mechanics.* 1967, VIII + 427 pp.
8. Zdeněk Kopal and Constantine L. Goudas (eds.), *Measure of the Moon. Proceedings of the 2nd International Conference on Selenodesy and Lunar Topography, held in the University of Manchester, England, May 30–June 4, 1966.* 1967, XVIII + 479 pp.
9. J. G. Emming (ed.), *Electromagnetic Radiation in Space. Proceedings of the 3rd ESRO Summer School in Space Physics, held in Alpbach, Austria, from 19 July to 13 August, 1965.* 1968, VIII + 307 pp.
10. R. L. Carovillano, John, F. McClay, and Henry R. Radoski (eds.), *Physics of the Magnetosphere, Based upon the Proceedings of the Conference held at Boston College, June 19–28, 1967.* 1968, X + 686 pp.
11. Syun-Ichi Akasofu, *Polar and Magnetospheric Substorms.* 1968, XVIII + 280 pp.
12. Peter M. Millman (ed.), *Meteorite Research. Proceedings of a Symposium on Meteorite Research, held in Vienna, Austria, 7–13 August, 1968.* 1969, XV + 941 pp.
13. Margherita Hack (ed.), *Mass Loss from Stars. Proceedings of the 2nd Trieste Colloquium on Astrophysics, 12–17 September, 1968.* 1969, XII + 345 pp.
14. N. D'Angelo (ed.), *Low-Frequency Waves and Irregularities in the Ionosphere. Proceedings of the 2nd ESRIN-ESLAB Symposium, held in Frascati, Italy, 23–27 September, 1968.* 1969, VII + 218 pp.
15. G. A. Partel (ed.), *Space Engineering. Proceedings of the 2nd International Conference on Space Engineering, held at the Fondazione Giorgio Cini, Isola di San Giorgio, Venice, Italy, May 7–10, 1969.* 1970, XI + 728 pp.
16. S. Fred Singer (ed.), *Manned Laboratories in Space. Second International Orbital Laboratory Symposium.* 1969, XIII + 133 pp.
17. B. M. McCormac (ed.), *Particles and Fields in the Magnetosphere. Symposium Organized by the Summer Advanced Study Institute, held at the University of California, Santa Barbara, Calif., August 4–15, 1969.* 1970, XI + 450 pp.
18. Jean-Claude Pecker, *Experimental Astronomy.* 1970, X + 105 pp.
19. V. Manno and D. E. Page (eds.), *Intercorrelated Satellite Observations related to Solar Events. Proceedings of the 3rd ESLAB/ESRIN Symposium held in Noordwijk, The Netherlands, September 16–19, 1969.* 1970, XVI + 627 pp.
20. L. Mansinha, D. E. Smylie, and A. E. Beck, *Earthquake Displacement Fields and the Rotation of the Earth, A NATO Advanced Study Institute Conference Organized by the Department of Geophysics, University of Western Ontario, London, Canada, June 22–28, 1969.* 1970, XI + 308 pp.
21. Jean-Claude Pecker, *Space Observatories.* 1970, XI + 120 pp.
22. L. N. Mavridis (ed.), *Structure and Evolution of the Galaxy. Proceedings of the NATO Advanced Study Institute, held in Athens, September 8–19, 1969.* 1971, VII + 312 pp.
23. A. Muller (ed.), *The Magellanic Clouds. A European Southern Observatory Presentation: Principal Prospects, Current Observational and Theoretical Approaches, and Prospects for Future Research, Based on the Symposium on the Magellanic Clouds, held in Santiago de Chile, March 1969, on the Occasion of the Dedication of the European Southern Observatory.* 1971, XII + 189 pp.

24. B. M. McCormac (ed.), *The Radiating Atmosphere. Proceedings of a Symposium Organized by the Summer Advanced Study Institute, held at Queen's University, Kingston, Ontario, August 3–14, 1970.* 1971, XI + 455 pp.
25. G. Fiocco (ed.), *Mesospheric Models and Related Experiments. Proceedings of the 4th ESRIN-ESLAB Symposium, held at Frascati, Italy, July 6–10, 1970.* 1971, VIII + 298 pp.
26. I. Atanasijević, *Selected Exercises in Galactic Astronomy.* 1971, XII + 144 pp.
27. C. J. Macris (ed.), *Physics of the Solar Corona. Proceedings of the NATO Advanced Study Institute on Physics of the Solar Corona, held at Cavouri-Vouliagmeni, Athens, Greece, 6–17 September 1970.* 1971, XII + 345 pp.
28. F. Delobeau, *The Environment of the Earth.* 1971, IX + 113 pp.
29. E. R. Dyer (general ed.), *Solar-Terrestrial Physics/1970. Proceedings of the International Symposium on Solar-Terrestrial Physics, held in Leningrad, U.S.S.R., 12–19 May 1970.* 1972, VIII + 938 pp.
30. V. Manno and J. Ring (eds.), *Infrared Detection Techniques for Space Research. Proceedings of the 5th ESLAB-ESRIN Symposium, held in Noordwijk, The Netherlands, June 8–11, 1971.* 1972, XII + 344 pp.
31. M. Lecar (ed.), *Gravitational N-Body Problem. Proceedings of IAU Colloquium No. 10, held in Cambridge, England, August 12–15, 1970.* 1972, XI + 441 pp.
32. B. M. McCormac (ed.), *Earth's Magnetospheric Processes. Proceedings of a Symposium Organized by the Summer Advanced Study Institute and Ninth ESRO Summer School, held in Cortina, Italy, August 30–September 10, 1971.* 1972, VIII + 417 pp.
33. Antonin Rükl, *Maps of Lunar Hemispheres.* 1972, V + 24 pp.
34. V. Kourganoff, *Introduction to the Physics of Stellar Interiors.* 1973, XI + 115 pp.
35. B. M. McCormac (ed.), *Physics and Chemistry of Upper Atmospheres. Proceedings of a Symposium Organized by the Summer Advanced Study Institute, held at the University of Orléans, France, July 31–August 11, 1972.* 1973, VIII + 389 pp.
36. J. D. Fernie (ed.), *Variable Stars in Globular Clusters and in Related Systems. Proceedings of the IAU Colloquium No. 21, held at the University of Toronto, Toronto, Canada, August 29–31, 1972.* 1973, IX + 234 pp.
37. R. J. L. Grard (ed.), *Photon and Particle Interaction with Surfaces in Space. Proceedings of the 6th ESLAB Symposium, held at Noordwijk, The Netherlands, 26–29 September, 1972.* 1973, XV + 577 pp.
38. Werner Israel (ed.), *Relativity, Astrophysics and Cosmology. Proceedings of the Summer School, held 14–26 August, 1972, at the BANFF Centre, BANFF, Alberta, Canada.* 1973, IX + 323 pp.
39. B. D. Tapley and V. Szebehely (eds.), *Recent Advances in Dynamical Astronomy. Proceedings of the NATO Advanced Study Institute in Dynamical Astronomy, held in Cortina d'Ampezzo, Italy, August 9–12, 1972.* 1973, XIII + 468 pp.
40. A. G. W. Cameron (ed.), *Cosmochemistry. Proceedings of the Symposium on Cosmochemistry, held at the Smithsonian Astrophysical Observatory, Cambridge, Mass., August 14–16, 1972.* 1973, X + 173 pp.
41. M. Golay, *Introduction to Astronomical Photometry.* 1974, IX + 364 pp.
42. D. E. Page (ed.), *Correlated Interplanetary and Magnetospheric Observations. Proceedings of the 7th ESLAB Symposium, held at Saulgau, W. Germany, 22–25 May, 1973.* 1974, XIV + 662 pp.
43. Riccardo Giacconi and Herbert Gursky (eds.), *X-Ray Astronomy.* 1974, X + 450 pp.
44. B. M. McCormac (ed.), *Magnetospheric Physics. Proceedings of the Advanced Summer Institute, held in Sheffield, U.K., August 1973.* 1974, VII + 399 pp.
45. C. B. Cosmovici (ed.), *Supernovae and Supernova Remnants. Proceedings of the International Conference on Supernovae, held in Lecce, Italy, May 7–11, 1973.* 1974, XVII + 387 pp.
46. A. P. Mitra, *Ionospheric Effects of Solar Flares.* 1974, XI + 294 pp.
47. S.-I. Akasofu, *Physics of Magnetospheric Substorms.* 1977, XVIII + 599 pp.
48. H. Gursky and R. Ruffini (eds.), *Neutron Stars, Black Holes and Binary X-Ray Sources.* 1975, XII + 441 pp.
49. Z. Švestka and P. Simon (eds.), *Catalog of Solar Particle Events 1955–1969. Prepared under the Auspices of Working Group 2 of the Inter-Union Commission on Solar-Terrestrial Physics.* 1975, IX + 428 pp.
50. Zdeněk Kopal and Robert W. Carder, *Mapping of the Moon.* 1974, VIII + 237 pp.
51. B. M. McCormac (ed.), *Atmospheres of Earth and the Planets. Proceedings of the Summer Advanced Study Institute, held at the University of Liège, Belgium, July 29–August 8, 1974.* 1975, VII + 454 pp.
52. V. Formisano (ed.), *The Magnetospheres of the Earth and Jupiter. Proceedings of the Neil Brice Memorial Symposium, held in Frascati, May 28–June 1, 1974.* 1975, XI + 485 pp.

53. R. Grant Athay, *The Solar Chromosphere and Corona: Quiet Sun.* 1976, XI + 504 pp.
54. C. de Jager and H. Nieuwenhuijzen (eds.), *Image Processing Techniques in Astronomy. Proceedings of a Conference, held in Utrecht on March 25–27, 1975,* XI + 418 pp.
55. N. C. Wickramasinghe and D. J. Morgan (eds.), *Solid State Astrophysics. Proceedings of a Symposium, held at the University College, Cardiff, Wales, 9–12 July 1974.* 1976, XII + 314 pp.
56. John Meaburn, *Detection and Spectrometry of Faint Light.* 1976, IX + 270 pp.
57. K. Knott and B. Battrick (eds.), *The Scientific Satellite Programme during the International Magnetospheric Study. Proceedings of the 10th ESLAB Symposium, held at Vienna, Austria, 10–13 June 1975.* 1976, XV + 464 pp.
58. B. M. McCormac (ed.), *Magnetospheric Particles and Fields. Proceedings of the Summer Advanced Study School, held in Graz, Austria, August 4–15, 1975.* 1976, VII + 331 pp.
59. B. S. P. Shen and M. Merker (eds.), *Spallation Nuclear Reactions and Their Applications.* 1976, VIII + 235 pp.
60. Walter S. Fitch (ed.), *Multiple Periodic Variable Stars. Proceedings of the International Astronomical Union Colloquium No. 29, Held at Budapest, Hungary, 1–5 September 1975.* 1976, XIV + 348 pp.
61. J. J. Burger, A. Pedersen, and B. Battrick (eds.), *Atmospheric Physics from Spacelab. Proceedings of the 11th ESLAB Symposium, Organized by the Space Science Department of the European Space Agency, held at Frascati, Italy, 11–14 May 1976.* 1976, XX + 409 pp.
62. J. Derral Mulholland (ed.), *Scientific Applications of Lunar Laser Ranging. Proceedings of a Symposium held in Austin, Tex., U.S.A., 8–10 June, 1976.* 1977, XVII + 302 pp.